Illuminate
Publishing

WJEC
AS Chemistry

Study and Revision Guide

Peter Blake • Elfed Charles • Kathryn Foster

Published in 2012 by Illuminate Publishing Ltd, P.O Box 1160,
Cheltenham, Gloucestershire GL50 9RW

Orders: Please visit www.illuminatepublishing.com
or email sales@illuminatepublishing.com

British Library Cataloguing in Publication Data

A catalogue record for this book is available from the British Library

ISBN 978-0-9568401-7-2

Printed by T J International, Padstow, Cornwall

The publisher's policy is to use papers that are natural, renewable and recyclable
products made from wood grown in sustainable forests. The logging and manufacturing
processes are expected to conform to the environmental regulations of the country of
origin.

This material has been endorsed by WJEC and offers high quality support for the
delivery of WJEC qualifications. While this material has been through a WJEC quality
assurance process, all responsibility for the content remains with the publisher.

Editor: Geoff Tuttle
Design: Nigel Harriss
Layout: Nigel Harriss and Claire Young

Acknowledgements
We are very grateful to the team at Illuminate Publishing for their professionalism,
support and guidance throughout this project. It has been a pleasure to work so closely
with them.

The author and publisher wish to thank:

Judith Bonello for her thorough review of the book and expert insights and observations.

We are indebted to Mike Ebbsworth of WJEC whose unstinting help and
encouragement from the beginning made this whole undertaking possible.

Jonathan Owen of WJEC.

Contents

How to use this book

As Principal Examiners for the WJEC specification we have written this study guide to help you be aware of what is required, and structured the content to guide you through to success in the WJEC Chemistry AS examination.

Knowledge and Understanding

The **first section** of the book covers the key knowledge and understanding required for the examination and provides notes for each of the two examination theory papers:

CH1 – Controlling and Using Chemical Changes

CH2 – Properties, Structure and Bonding

In addition, we have tried to give you additional pointers so that you can develop your work:

- Questions may be based on any term in the WJEC specification so these terms are defined and highlighted.

- There are 'Quickfire' questions designed to test your knowledge and understanding of the material.

- There is a comprehensive set of candidate answers to questions in all sections along with marking, analysis and explanation by the examiners of these answers.

- 'Grade Boost' inserts point out key ways in which candidates can impress the examiners by their knowledge and understanding.

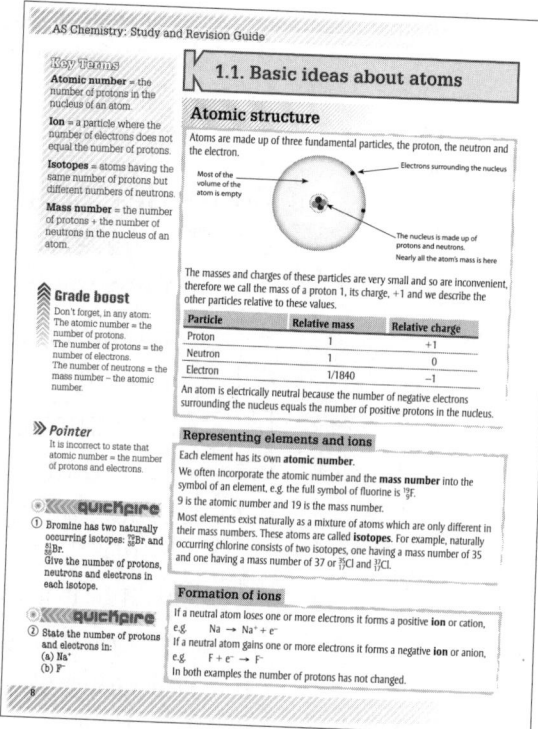

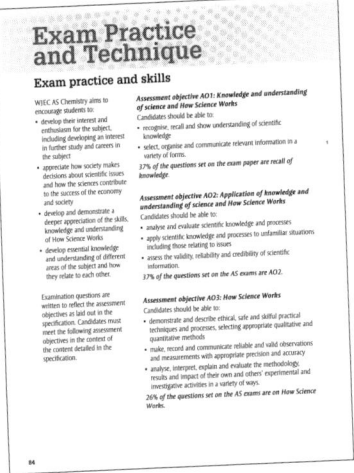

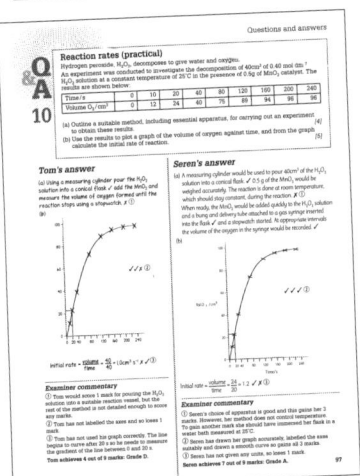

Exam Practice and Technique

The **second section** of the book covers the key skills for examination success and offers you examples based on real-life responses to examination questions. First you will be guided into an understanding of how the examination system works, and then offered clues to success.

A variety of structured and essay questions are provided in this section. Each essay includes the marking points expected followed by actual samples of candidates' responses. A variety of structured questions are also provided, together with typical responses and comments. They offer a guide as to the standard that is required, and the commentary will explain why the responses gained the marks that they did.

Most importantly, we advise that you should take responsibility for your own learning and not rely on your teachers to give you notes or tell you how to gain the grades that you require. You should look for additional notes to support your study into WJEC Chemistry.

We advise that you look at the WJEC website www.wjec.co.uk. In particular, you need to be aware of the specification. Look for specimen examination papers and mark schemes. You may find past papers useful as well.

Good luck with your revision.

Peter Blake, Elfed Charles and Kathryn Foster

Knowledge and Understanding

CH1 Controlling and Using Chemical Changes

This unit begins with some important fundamental ideas about atoms and the use of the mole concept in calculations.

Three key principles governing chemical change are then studied: the position of equilibrium between reactants and products, the energy changes associated with a chemical reaction and the rate at which reactions take place.

These principles are then applied to some important problems in the fields of chemical synthesis, obtaining energy and the maintenance of the environment.

Revision checklist

Tick column 1 when you have completed brief revision notes.
Tick column 2 when you think you have a good grasp of the topic.
Tick column 3 during final revision when you feel you have mastery of the topic.

Key Terms

Atomic number = the number of protons in the nucleus of an atom.

Ion = a particle where the number of electrons does not equal the number of protons.

Isotopes = atoms having the same number of protons but different numbers of neutrons.

Mass number = the number of protons + the number of neutrons in the nucleus of an atom.

Grade boost

Don't forget, in any atom:
The atomic number = the number of protons.
The number of protons = the number of electrons.
The number of neutrons = the mass number – the atomic number.

≫ Pointer

It is incorrect to state that atomic number = the number of protons and electrons.

① Bromine has two naturally occurring isotopes: $^{79}_{35}$Br and $^{81}_{35}$Br.
Give the number of protons, neutrons and electrons in each isotope.

② State the number of protons and electrons in:
(a) Na⁺
(b) F⁻

1.1. Basic ideas about atoms

Atomic structure

Atoms are made up of three fundamental particles, the proton, the neutron and the electron.

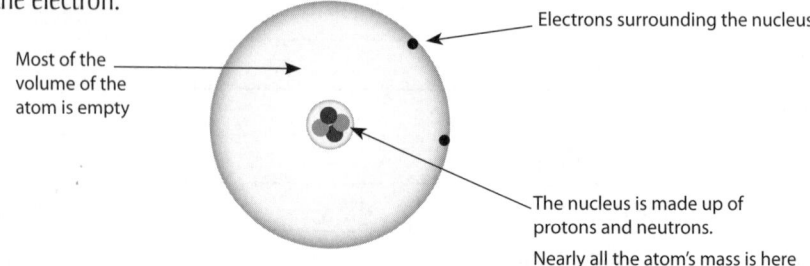

Most of the volume of the atom is empty

Electrons surrounding the nucleus

The nucleus is made up of protons and neutrons.
Nearly all the atom's mass is here

The masses and charges of these particles are very small and so are inconvenient, therefore we call the mass of a proton 1, its charge, +1 and we describe the other particles relative to these values.

Particle	Relative mass	Relative charge
Proton	1	+1
Neutron	1	0
Electron	1/1840	−1

An atom is electrically neutral because the number of negative electrons surrounding the nucleus equals the number of positive protons in the nucleus.

Representing elements and ions

Each element has its own **atomic number**.

We often incorporate the atomic number and the **mass number** into the symbol of an element, e.g. the full symbol of fluorine is $^{19}_{9}$F.

9 is the atomic number and 19 is the mass number.

Most elements exist naturally as a mixture of atoms which are only different in their mass numbers. These atoms are called **isotopes**. For example, naturally occurring chlorine consists of two isotopes, one having a mass number of 35 and one having a mass number of 37 or $^{35}_{17}$Cl and $^{37}_{17}$Cl.

Formation of ions

If a neutral atom loses one or more electrons it forms a positive **ion** or cation,
e.g. $Na \rightarrow Na^+ + e^-$
If a neutral atom gains one or more electrons it forms a negative **ion** or anion,
e.g. $F + e^- \rightarrow F^-$
In both examples the number of protons has not changed.

Radioactivity

Types of radioactive emission and their behaviour

Some isotopes are unstable. Their nuclei spontaneously disintegrate and emit either **alpha (α) particles** or **beta (β) particles** but never both. **Gamma (γ) rays** may also be given off.

Radiation	Nature	Effect of electric field	Penetrating power
α particles	clusters of 2 protons and 2 neutrons	attracted to negative plate	least penetrating stopped by a piece of paper
β particles	electrons	attracted to positive plate	stopped by a thin sheet of metal (e.g. 0.5 cm of aluminium)
γ rays	high energy electromagnetic radiation	no effect	most penetrating may take more than 2 cm of lead to stop them

Effect on mass number and atomic number

α and β particle emissions result in the formation of a new nucleus with a new atomic number therefore the product is a different element.

When an element emits an α particle its mass number decreases by 4 and its atomic number decreases by 2.

$$^{238}_{92}U \rightarrow {}^{234}_{90}Th + {}^{4}_{2}\alpha$$

The product is two places to the left in the Periodic Table.

When an element emits a β particle its mass number is unchanged and its atomic number increases by 1.

$$^{14}_{6}C \rightarrow {}^{14}_{7}N + {}_{-1}\beta$$

The product is one place to the right in the Periodic Table.

Key Terms

α particles = cluster of 2 protons and 2 neutrons, therefore positively charged.

β particles = fast moving electrons, therefore negatively charged.

γ rays = high energy electromagnetic radiation, therefore no charge.

Grade boost

β particles can be considered as being formed when a neutron changes into a proton,
i.e. $^{1}_{0}n \rightarrow {}^{1}_{1}p + {}_{-1}\beta$

Pointer

In equations:
$^{4}_{2}He^{2+}$ is acceptable for $^{4}_{2}\alpha$
$^{0}_{-1}e$ is acceptable for $_{-1}\beta$.

quickfire

③ Why are victims of radioactive contamination buried in lead coffins?

quickfire

④ Give the mass number and symbol of the isotope formed when ^{211}Bi decays by α emission.

Key Term

Half-life = the time taken for half the atoms in a radioisotope to decay or the time taken for the radioactivity of a radioisotope to fall to half its initial value.

Consequences for living cells

Radioactive emissions are potentially harmful. However, we all receive some radiation from the normal background radiation that occurs everywhere. Workers in industries where they are exposed to radiation from radioactive isotopes are carefully monitored to ensure that they do not receive more radiation than is allowed under internationally agreed limits.

High energy radioactive emissions break chemical bonds in the cell molecules giving rise to changes in DNA which can cause mutations and the formation of cancerous cells at lower doses, or cell death at higher doses.

When α particle emitting isotopes are ingested they are far more dangerous than an equivalent activity of β emitting or γ emitting isotopes but fortunately α particles from outside cannot penetrate the skin.

quickfire

⑤ Outline why radioactivity may be a health hazard.

》 Pointer

The greater the half-life of a radioactive isotope, the greater the concern, since the radioactivity of the isotope exists for a longer time.

quickfire

⑥ An isotope of iodine ^{131}I has a half-life of 8 days. Calculate how long it would take for 1.00 g of ^{131}I to be reduced to 0.125 g of ^{131}I.

Half-life

In calculations involving **half-life**, the half-life of radioactive isotopes will always be given. Two types of calculation can be set:

- Finding the time taken for the radioactivity of a sample to fall to a certain fraction of its initial value.
- Finding the mass of a radioactive isotope remaining after a certain length of time given the initial mass.

Example

Strontium-90 has a half-life of 27 years.

(a) Calculate how long it will take for the activity of the isotope to decay to $1/16^{th}$ of its original value.

(b) Calculate the mass remaining after 81 years if 1.6 grams are originally present.

(a)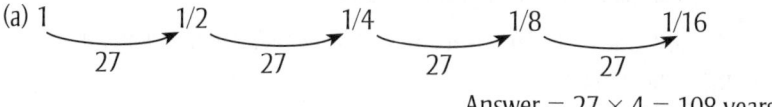

Answer = $27 \times 4 = 108$ years

(b) 81 years = 3 half-lives

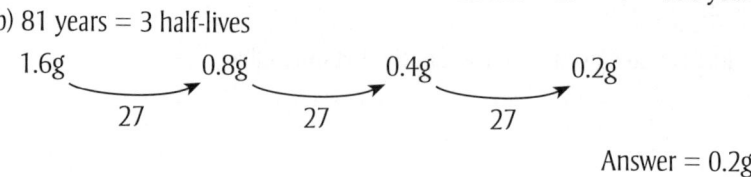

Answer = 0.2g

Beneficial uses of radioactivity

Below are some examples. Candidates should be able to give an example in each area.

Medicine

- Cobalt-60 in radiotherapy for the treatment of cancer. The high energy of γ radiation is used to kill cancer cells and prevent the malignant tumour from developing.
- Iodine-131 for patients with defective thyroid glands. The iodine-131 acts as a tracer to study the uptake of iodine in the gland.

Radio-dating

- Carbon-14 (half-life 5570 years) is used to calculate the age of plant and animal remains. All living organisms absorb carbon, which includes a small proportion of the radioactive carbon-14. When an organism dies there is no more absorption of carbon-14 and that which is already present decays. The rate of decay decreases over the years and the activity that remains can be used to calculate the age of organisms.
- Potassium-40 (half-life 1300 million years) is used to estimate the geological age of rocks. Potassium-40 can change into argon-40 by the nucleus gaining an inner electron. Measuring the ratio of potassium-40 to argon-40 in a rock gives an estimate of its age.

Analysis

- Dilution analysis. The use of isotopically labelled substances to find the mass of a substance in a mixture. This is useful when a component of a complex mixture can be isolated from the mixture in the pure state but cannot be extracted quantitatively.
- Monitoring the thickness of metal strips or foil. The metal is placed between two rollers to get the right thickness. A radioactive source (a β emitter) is mounted on one side of the metal with a detector on the other. If the amount of radiation reaching the detector increases, the detector operates a mechanism for moving the rollers apart and vice versa.

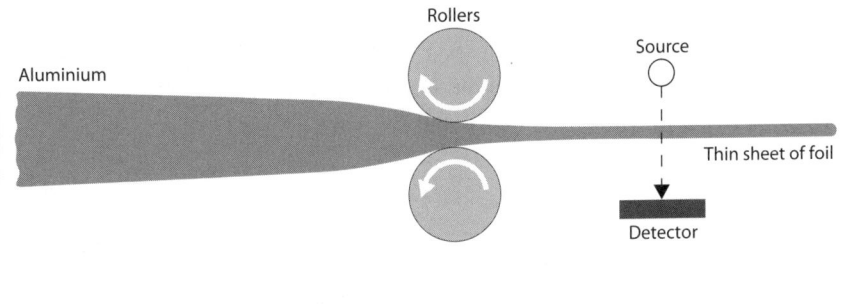

quickfire

⑦ Radioactive iodine, ^{131}I is used in medicine as a tracer. Give another use of radioactive isotopes apart from in medicine.

» Pointer

Other uses of radioactive isotopes include detection of leaks in water or fuel pipes, measuring engine wear in cars, structure determination and studies of reaction mechanisms.

Key Term

Atomic orbital = a region in an atom that can hold up to two electrons with opposite spins.

≫ Pointer

The s sub-shell can hold 2 electrons.
The p sub-sh ell can hold 6 electrons.
The d sub-shell can hold 10 electrons.

▲ Grade boost

Sodium is classified as an s-block element because its outer electron is in an s orbital. Chlorine is classified as a p-block element because its outer electron is in a p orbital.

Electron shells or energy levels

Electrons within atoms occupy fixed energy levels or shells. Each shell has a principal quantum number, n. The lower the value of n, the closer the shell to the nucleus and the lower the energy level.

Electron sub-shells or orbitals

In a shell there are regions of space around the nucleus where there is a high probability of finding an electron of a given energy. These regions are called **atomic orbitals**. Each orbital can contain two electrons. Electrons have a property called spin. In order for two electrons, both of which have a negative charge, to exist in the same orbital they must have opposite spins. There are four different types of orbital: s, p, d, and f. Orbitals of the same type are grouped together as a sub-shell.

An s orbital is spherical and can contain two electrons.

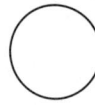

The boundary represents an area where the electron spends 90% of its existence.

A p orbital is made up of three dumb-bell shaped lobes mutually at right angles. They are shown separated below:

 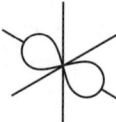 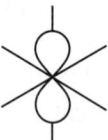

p_x orbital $\qquad$ p_y orbital $\qquad$ p_z orbital

Since each p orbital can hold two electrons, a p sub-shell can hold 6 electrons in total.

There are five d orbitals; therefore there are a total of 10 electrons in a d sub-shell.

This is how the Periodic Table looks in terms of s, p and d electrons.

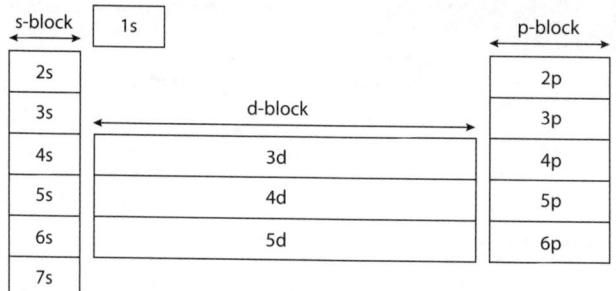

The f-block elements have been omitted as they are not needed for the exams.

Filling orbitals with electrons

The way in which electrons are arranged can be worked out using three basic rules:

1. Electrons fill atomic orbitals in order of increasing energy.

2. A maximum of two electrons can occupy any orbital each with opposite spins.

3. Each orbital in a sub-shell will first fill with one electron before pairing starts.

Electronic structures or **configurations** can be represented by 'electrons in boxes'. Each orbital is represented as a box and the electrons as arrows in the boxes. The opposite spin of paired electrons is shown by arrows facing up and down.

e.g. Nitrogen has 7 electrons

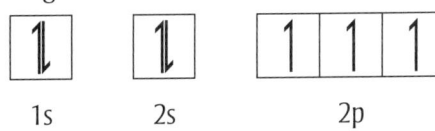

A shorter way of showing electronic structure is to write the shell number first, followed by the orbital letter and then the number of electrons in the orbital which are written as a superscript.

For example, since nitrogen has:

2 electrons in the s orbital in the 1st shell

2 electrons in the s orbital in the 2nd shell

3 electrons in the p orbital in the 2nd shell,

its electronic configuration is $1s^2 2s^2 2p^3$.

Calcium has 20 electrons, it electronic configuration is : $1s^2 2s^2 2p^6 3s^2 3p^6 4s^2$.

The electronic configuration of ions is presented in the same way as that of atoms.

Positive ions form by the loss of electrons from the highest energy orbitals so these ions have fewer electrons than the parent atom.

Negative ions form by adding electrons to the highest energy orbitals so these ions have more electrons than the parent atom.

e.g. Na $1s^2 2s^2 2p^6 3s^1$ Na$^+$ $1s^2 2s^2 2p^6$
Cl $1s^2 2s^2 2p^6 3s^2 3p^5$ Cl$^-$ $1s^2 2s^2 2p^6 3s^2 3p^6$

Key Term

Electronic configuration
= the arrangement of electrons in an atom.

 Grade boost

The 4s orbitals are filled before the 3d orbitals.

 Grade boost

The configurations for chromium and copper are not as expected, they both end in $4s^1$.

» Pointer

You need to know the electronic configuration for the first 36 elements but only for the ions of the first 20 elements.

quickfire

⑧ (a) Use electrons in boxes to write the electronic configuration of:
(i) an atom of phosphorus, P
(ii) a sulfide ion, S^{2-}.

(b) Write the electronic configuration in terms of sub-shells for a copper atom.

» Pointer

If the conditions for ionisation energy are 298 K and 1 atm then the process is known as the standard ionisation energy.

quickfire

⑨ State and explain the general trend in ionisation energy:
(a) across a period
(b) down a group.

quickfire

⑩ Write an equation to represent the second ionisation energy of magnesium.

Ionisation energies

The process of removing electrons from an atom is called ionisation. The energy needed to remove each successive electron from an atom is called the first, second, etc., ionisation energy.

The process for the first ionisation energy (IE) of an element is summarised in the equation:

$$X(g) \rightarrow X^+(g) + e^-$$

Electrons are held in their shells by their attraction to the positive nucleus, therefore the greater the attraction, the greater the ionisation energy. This attraction depends on three factors:

- Nuclear charge – the greater the nuclear charge, the greater the attractive force on the outer electron.
- **Electron shielding** – the more inner shells or sub-shells there are, the smaller the attractive force on the outer electron.
- Atomic radius – the greater the atomic radius, the smaller the attractive force on the outer electron.

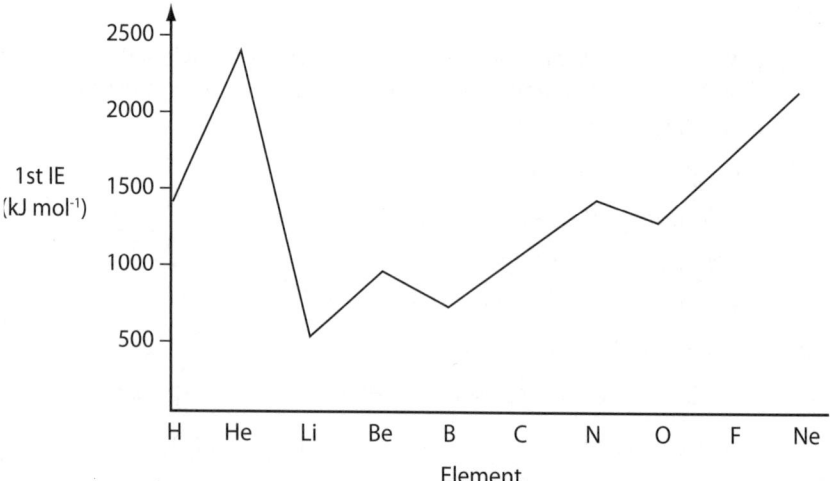

A plot of first IE against the first ten elements of the Periodic Table shows evidence of shells and sub-shells:

He > Ne since neon's outer electron has increased shielding from inner electrons and is further from the nucleus.

He > Li since lithium's outer electron is in a new shell which has increased shielding and is further from the nucleus.

Ne > O effect of greater nuclear charge in same sub-shell; little extra shielding.

Be > B since boron's outer electron is in a new sub-shell further from the nucleus and is partly shielded by the 2s electrons.

N > O since the electron-electron repulsion between the two paired electrons in one p orbital in oxygen makes one of the electrons easier to remove. Nitrogen does not contain paired electrons.

Successive ionisation energies

Successive ionisation energies are a measure of the energy needed to remove each electron in turn until all the electrons are removed from an atom.

An element has as many ionisation energies as it has electrons. Sodium has eleven electrons and so has eleven successive ionisation energies.

For example, the third ionisation energy is a measure of how easily a 2^+ ion loses an electron to form a 3^+ ion. An equation to represent the third ionisation energy of sodium is:

$$Na^{2+}(g) \rightarrow Na^{3+}(g) + e^-$$

Successive ionisation energies always increase because:

- As each electron is removed there is less electron – electron repulsion and each shell will be drawn in slightly closer to the nucleus
- As the distance of each electron from the nucleus decreases, the nuclear attraction increases.

The graph below shows how the successive ionisation energies of sodium provide further evidence for the existence of different shells.

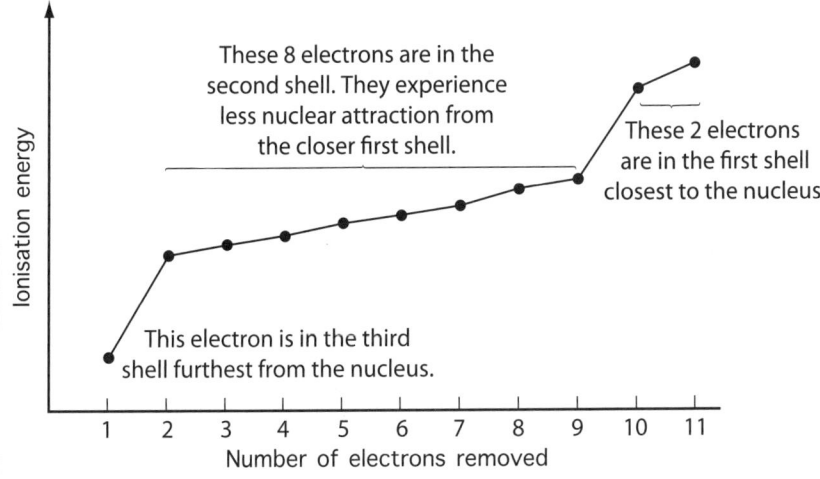

For sodium there is one electron on its own which is easiest to remove. Then there are eight more electrons which become successively more difficult to remove. Finally there are two electrons which are the most difficult to remove.

Notice the large increases in ionisation energy as the 2nd and 10th electrons are removed. If the electrons were all in the same shell, there would be no large rise or jump.

Grade boost

A large increase in successive ionisation energies shows that an electron has been removed from a new shell closer to the nucleus and gives the group to which the element belongs.
Li has a large energy jump between 1st and 2nd IE therefore it's in group 1.
Al has a large energy jump between 3rd and 4th IE therefore it's in group 3.

 quickfire

⑪ The first four ionisation energies (in kJ mol^{-1}) for an element are: 738, 1451, 7733 and 10541.
State and explain to which group in the Periodic Table the element belongs.

Emission and absorption spectra

Energy, frequency and wavelength

» Pointer

Light is electromagnetic radiation in the range of wavelength corresponding to the visible region of the electromagnetic spectrum.

» Pointer

The electromagnetic spectrum is the range of all possible frequencies of electromagnetic radiation.

Light is a form of electromagnetic radiation. The frequency and wavelength of light are related by the equation:

$$c = f\lambda \quad \text{(c is the speed of light)}$$

The frequency of electromagnetic radiation and energy are connected by the equation:

$$E = hf \quad \text{(h is Planck's constant)}$$

Therefore, $f \propto E$, (f is directly proportional to E) and if frequency increases energy increases.

$f \propto 1/\lambda$ (f is inversely proportional to λ) and if frequency increases wavelength decreases.

⏶ Grade boost

Since $f \propto E$ and $f \propto 1/\lambda$, the lower the wavelength, the higher the frequency and the greater the energy.

Absorption spectra

When white light shines through gaseous atoms, photons of certain energy may be absorbed by an atom causing an electron to move from a lower energy level to a higher one. This means that light of a frequency corresponding to the energy of the photon will be removed. Therefore when the light is examined by a spectrometer, the electronic transitions will appear as dark lines against a bright background.

⊙ «« quicKfire

(12) Two lines in the emission spectrum of atomic hydrogen have the following frequencies 460 THz and 690 THz. State and explain which one has the higher:
(a) energy?
(b) wavelength?

Emission spectra

When atoms are given energy by heating or by an electrical field, electrons are promoted from a lower energy level to a higher one. When the source of energy is removed, the electrons fall from the higher energy level to a lower energy level and the energy lost is released as a packet of energy called a quantum of energy. This corresponds to electromagnetic radiation of a specific frequency. The observed spectrum consists of a number of coloured lines on a black background:

⊙ «« quicKfire

(13) State briefly the difference between absorption and emission spectra.

If the electron energy levels were not quantised but could have any value, a continuous spectrum rather than a line spectrum would result:

The hydrogen spectrum

An atom of hydrogen has only one electron so it gives the simplest emission spectrum. The atomic spectrum of hydrogen consists of separate series of lines mainly in the ultraviolet, visible and infrared regions of the electromagnetic spectrum. Only one series, the Balmer series, is in the visible region.

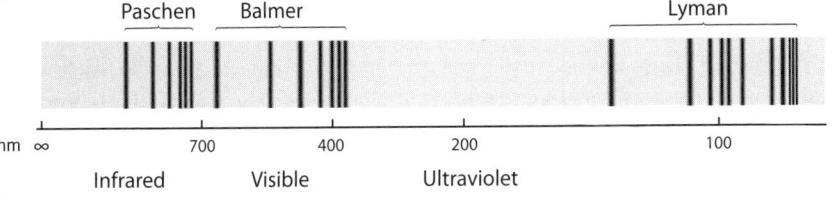

Part of emission spectrum of atomic hydrogen

When an atom is excited by absorbing energy, an electron jumps up to a higher energy level. As the electron falls back down to a lower energy level, it emits energy in the form of electromagnetic radiation. The emitted energy can be seen as a line in the spectrum because the energy of the emitted radiation is equal to the difference between the two energy levels, ΔE, in this electronic transition, i.e. it is a fixed quantity or quantum.

Since $\Delta E = hf$, electronic transitions between different energy levels result in emission of radiation of different frequencies and therefore produce different lines in the spectrum.

Each line in the Balmer series is due to electrons returning to the second shell or $n = 2$ energy level. As the frequency increases, the lines get closer together because the energy difference between the shells decreases.

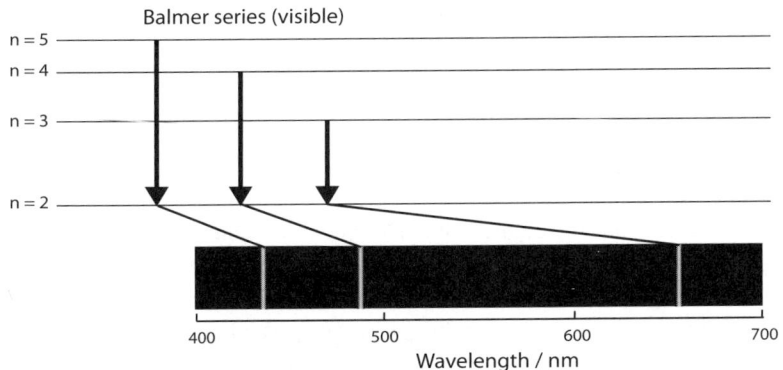

Ionisation of the hydrogen atom

The spectral lines become closer and closer together as the frequency of the radiation increases until they converge to a limit. The convergence limit corresponds to the point at which the energy of an electron is no longer quantised. At that point the nucleus has lost all influence over the electron; the atom has become ionised.

Measuring the convergent frequency of the Lyman series (difference from $n = 1$ to $n = \infty$) allows the ionisation energy to be calculated using $\Delta E = hf$.

> **Pointer**
> Electronic transition is when an electron moves from one energy level to another.

> **Pointer**
> In the Balmer series electrons return to energy level $n = 2$. In the Lyman series electrons return to energy level $n = 1$.

> **Pointer**
> The convergence limit is when the spectral lines become so close together they have a continuous band of radiation and separate lines cannot be distinguished.

> **Grade boost**
> The IE of a hydrogen atom can be shown on its electron energy level diagram by drawing an arrow upwards from the $n = 1$ to the $n = \infty$ level.

> **quickfire**
> (14) In the atomic spectrum of hydrogen, why do the spectral lines get closer at the high frequency end of the Balmer series?

quickfire

⑮ What is the relative molecular mass, M_r, of:
(a) NH_4NO_3?
(b) $CuSO_4.5H_2O$?

≫ Pointer

Molar mass is the mass of one mole of a substance.

1.2 Chemical calculations

Atomic and molecular masses

Masses of atoms

The masses of atoms are too small to be used in calculations in chemical reactions, so instead the mass of an atom is expressed relative to a chosen standard atomic mass. The carbon-12 isotope is taken as the standard of reference.

Most elements exist naturally as two or more different isotopes. The mass of an element therefore depends on the relative abundance of all the isotopes present in the sample. In order to overcome this, chemists use an average mass of all the atoms and this is called the **relative atomic mass, A_r**.

Relative atomic mass has no units since it is one mass compared to another mass.

If we refer to the mass of a particular isotope then the term **relative isotopic mass** is used.

Masses of molecules

Many elements and compounds are made up of simple molecules, e.g. O_2, H_2O. The mass of a molecule is measured as the **relative molecular mass, M_r**, and is the sum of the relative atomic masses of all the atoms present in one molecule.

For ionic compounds, the formula represents a formula unit, rather than a molecule of the compound. Strictly speaking we should refer to relative formula mass not relative molecular mass.

Amount of substance

In chemical reactions the atoms that make up the reactants rearrange to form the products. For all the reactants to change into products the correct quantity of each reactant must be used. Since atoms are too small to be counted individually, chemists count atoms by weighing a collection of them where the mass of a particular fixed number of atoms is known.

Again carbon-12 is chosen as the standard and the number of atoms in exactly 12 g of carbon-12 is called a **mole**. This is a large number, 6.02×10^{23}, and is called the Avogadro constant, L.

When using mole to describe the amount of substance it is important to state the particles to which it refers. One mole of oxygen atoms is different to one mole of oxygen molecules.

The mass per mole of an element or compound is called the **molar mass, M**. It has the same numerical value as A_r or M_r but has the unit g mol^{-1}.

The mass spectrometer

In order to calculate the average mass of an atom of an element, the mass of the isotopes of the element together with their relative abundances must be known. These values are found using a mass spectrometer.

The processes in a mass spectrometer may be summarised as:

- Ionisation – a gaseous sample is bombarded with high energy electrons to form positive ions.
- Acceleration – an electric field accelerates the positive ions to high speed.
- Deflection – a magnetic field deflects the ions according to their mass/charge ratio. (Heavier ions are deflected less than light ones.)
- Detection – ions with the correct mass/charge ratio pass through a slit and are detected by an instrument such as an electrometer. The signal is then amplified and recorded.

The processes take place under high vacuum to prevent collision with air molecules.

》 Pointer

You do not need to be able to draw a diagram of a mass spectrometer but you could be required to label a diagram of one.

quickfire

⑯ State why the space inside a mass spectrometer is connected to a vacuum pump.

Calculating relative atomic masses

Below is the mass spectrum of magnesium

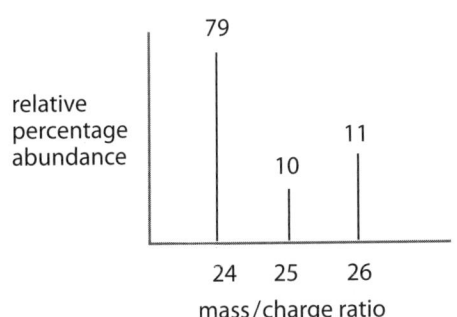

relative percentage abundance

mass/charge ratio

There are three peaks in the spectrum, so magnesium has three isotopes. The heights of the peaks give the relative abundances of the isotopes, which are given as percentages in the spectrum.

The relative atomic mass is a weighted average of the masses of all the atoms in the isotopic mixture, therefore:

$$\text{Relative atomic mass} = \frac{(79 \times 24) + (10 \times 25) + (11 \times 26)}{100} = 24.32$$

Other uses of mass spectrometry include:

- identifying unknown compounds
- identifying trace compounds in forensic science
- analysing molecules in space.

quickfire

⑰ A sample of strontium showed three peaks. The first peak was at m/z 86 and had an abundance of 10%; the second peak was at m/z 87 and had an abundance of 7%; the third peak was at m/z 88 and had an abundance of 83%. Calculate the relative atomic mass to three significant figures.

The mass spectrum of chlorine

Chlorine is made up of two isotopes ^{35}Cl and ^{37}Cl. However, chlorine gas consists of molecules not individual atoms and the mass spectrum of chlorine is:

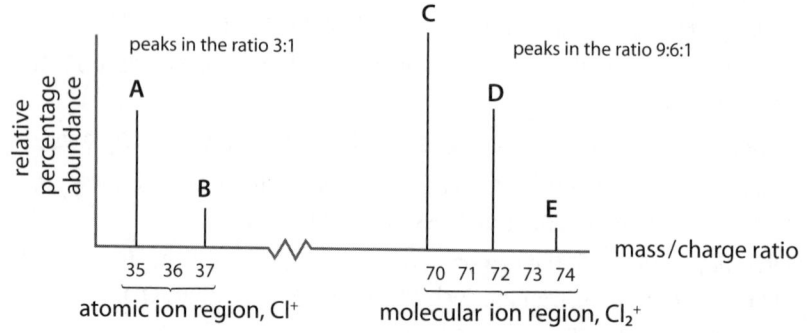

Mass spectrum of chlorine

>> **Pointer**

A molecular ion, M^+, is the positive ion formed in mass spectrometry when a molecule loses an electron.

>> **Pointer**

For the chlorine spectrum you can't make any predictions about the relative heights of the lines at m/z 35/37 compared with those at 70/72/74. That depends on what proportion of the molecular ions break up into fragments.

When chlorine is passed into the ionisation chamber, an electron is knocked off the molecule to give a molecular ion, Cl_2^+. These ions won't be particularly stable, and some will fall apart to give a chlorine atom and a Cl^+ ion. (This is known as fragmentation.)

So peak A is caused by $^{35}Cl^+$ and peak B by $^{37}Cl^+$.

As the ^{35}Cl isotope is three times more common than the ^{37}Cl isotope, the heights of the peaks are in the ratio of 3:1.

In the molecular ion region think about the possible combinations of ^{35}Cl and ^{37}Cl atoms in a Cl_2^+ ion. Both atoms could be ^{35}Cl, both atoms could be ^{37}Cl, or you could have one of each sort.

So peak C (m/z 70) is due to $(^{35}Cl\text{-}^{35}Cl)^+$

Peak D (m/z 72) is due to $(^{35}Cl\text{-}^{37}Cl)^+$ or $(^{37}Cl\text{-}^{35}Cl)^+$

Peak E (m/z 74) is due to $(^{37}Cl\text{-}^{37}Cl)^+$

Since the probability of an atom being ^{35}Cl is $3/4$ and that of being ^{37}Cl is $1/4$, then

molecule	$^{35}Cl\text{-}^{35}Cl$	$^{35}Cl\text{-}^{37}Cl$ or $^{37}Cl\text{-}^{35}Cl$	$^{37}Cl\text{-}^{37}Cl$
probability	$3/4 \times 3/4$	$3/4 \times 1/4$ or $1/4 \times 3/4$	$1/4 \times 1/4$
	$9/16$	$6/16$	$1/16$

and ratio of peaks C : D : E is 9 : 6 : 1

◉✕✕✕ **quickƒire**

⑱ In the mass spectrum of chlorine, explain why peaks due to chlorine atoms are present although chlorine gas contains only Cl_2 molecules.

Empirical and molecular formulae

Empirical formula is the simplest formula showing the simplest whole number ratio of the amount of elements present.

Molecular formula shows the actual number of atoms of each element present in the molecule. It is a simple multiple of the empirical formula. Usually the relative molecular mass is needed to determine the molecular formula.

Example

A compound of carbon, hydrogen and oxygen has a relative molecular mass of 180. The percentage composition by mass is C 40.0%; H 6.70%; O 53.3%. What is (a) the empirical formula and (b) the molecular formula?

(a)

	C :	H :	O
Molar ratio of atoms	$\dfrac{40}{12}$	$\dfrac{6.7}{1.01}$	$\dfrac{53.3}{16}$
=	3.33	6.63	3.33
Divide by smallest number	1	2	1
Empirical formula is		CH_2O	

(b) Mass of empirical formula $= 12 + 2.02 + 16 = 30.02$

Number of CH_2O units in a molecule $= \dfrac{180}{30.02} = 6$

Molecular formula is $C_6H_{12}O_6$

Moles, mass and concentration

The amount of substance and mass are linked by the equation:

$$\text{number of moles (n)} = \frac{\text{mass of substance (m)}}{\text{molar mass (M)}}$$

Concentration of a solution tells you how much solute is dissolved in the solvent. Its unit is normally mol dm^{-3}. The amount of substance and concentration are linked by the equation:

$$\text{number of moles (n)} = \text{concentration (c)} \times \text{volume (v)}$$

In the laboratory, volumes are normally measured in cm^3, so they have to be divided by 1000 to change them into dm^3.

Example

What mass of sodium carbonate is needed to prepare 250 cm^3 of a 0.400 mol dm^{-3} solution?

Calculate amount of Na_2CO_3 in moles:

$$n = c \times v = 0.400 \times 250/1000 = 0.100$$

Convert moles into mass: $M(Na_2CO_3) = 46 + 12 + 48 = 106$ g mol^{-1}

$$m = n \times M = 0.1 \times 106 = 10.6 \text{ g}$$

Grade boost

Learn the equations that connect amount of substance and mass for a solid:

$$n = \frac{m}{M}$$

$$or \quad m = nM$$

$$or \quad M = \frac{m}{n}$$

Grade boost

Learn the equations that connect amount of substance and concentration for a solution:

$$n = cv$$

$$or \quad c = \frac{n}{v}$$

$$or \quad v = \frac{n}{c}$$

Remember if v is given in cm^3 divide by a 1000 to change it into dm^3.

quickfire

⑲ Calculate the mass of 0.50 mol of calcium hydroxide $Ca(OH)_2$.

quickfire

⑳ Calculate the concentration of 4.0 g sodium hydroxide in 500 cm^3 of solution.

Equations for the reactions of solids and gases

An equation tells us not only what substances react together but also what amounts of substances react together. The number of moles of substances, as given by the balanced equation, are called the stoichiometric amounts.

The number of moles of solids can be calculated from their masses but the number of moles of gases are calculated from their volumes using the molar gas volume.

Example

In the presence of a catalyst, potassium chlorate(V) decomposes on heating to give potassium chloride and oxygen according to the equation:

$$2KClO_3(s) \rightarrow 2KCl(s) + 3O_2(g)$$

1.226 g of $KClO_3$ is heated until it is fully decomposed.

Calculate (a) the mass of KCl and (b) the volume of oxygen produced at 0°C and 1 atm

(1 mole of oxygen occupies 22.4 dm^3 at 0°C and 1 atm).

(a) Step 1. Calculate the amount, in mol, of 1.226 g $KClO_3$

$$n = \frac{m}{M} = \frac{1.226}{122.6} = 0.0100$$

Step 2. Use the equation to calculate the amount, in mol, of KCl formed

2 mol $KClO_3$ gives 2 mol KCl

0.01 mol $KClO_3$ gives 0.01 mol KCl

Step 3. Calculate the mass of KCl

Mass KCl = nM = 0.01 × 74.6 = 0.746 g

(b) Step 1. Amount of $KClO_3$ = 0.0100

Step 2. Calculate the amount of O_2 produced

2 mol $KClO_3$ gives 3 mol O_2

0.01 mol $KClO_3$ gives 0.015 mol O_2

Step 3. Calculate the volume of O_2 that is produced.

Volume O_2 = n × 22.4 = 0.015 × 22.4 = 0.336 dm^3 or 336 cm^3.

Atom economy and percentage yield

When a reaction occurs, the compounds formed, other than the product needed, are a waste. An indication of the efficiency of a reaction can be given as its **atom economy** or its **percentage yield**. The higher the atom economy, the more efficient the process.

Example

Benzenecarboxylic acid, C_6H_5COOH, can be formed from ethylbenzenecarboxylate, $C_6H_5COOC_2H_5$, according to the equation:

$$C_6H_5COOC_2H_5 + H_2O \rightarrow C_6H_5COOH + C_2H_5OH$$

If 1.95 g of acid was obtained from 3.00 g of ethylbenzenecarboxylate, calculate:

(a) the percentage yield

(b) the atom economy for the reaction

(a) % yield $= \dfrac{\text{mass of product obtained}}{\text{maximum theoretical mass}} \times 100\%$

To calculate the maximum mass the 3 steps in the previous example must be followed:

Step 1. Calculate the amount, in mol, of 3.00 g $C_6H_5COOC_2H_5$

$$n = \frac{m}{M} = \frac{3.00}{150.1} = 0.0200$$

Step 2. Use the equation to calculate the amount, in mol, of C_6H_5COOH formed

1 mol $C_6H_5COOC_2H_5$ gives 1 mol C_6H_5COOH

0.02 mol $C_6H_5COOC_2H_5$ gives 0.02 mol C_6H_5COOH

Step 3. Calculate the mass of C_6H_5COOH

$m = nM = 0.02 \times 122.06 = 2.44$ g

% yield $= \dfrac{1.95}{2.44} \times 100 = 79.9\%$

(b) Atom economy $= \dfrac{\text{mass of required product}}{\text{total mass of reactants}} \times 100\%$

$= \dfrac{M(C_6H_5COOH)}{M(C_6H_5COOC_2H_5) + M(H_2O)} \times 100 = \dfrac{122.06}{(150.1 + 18.02)} \times 100 = 72.6\%$

≫ Pointer

A reversible reaction is one which takes place in both forward and backward directions.

quickfire

① Explain the term *dynamic equilibrium* for a chemical system.

2.1 Chemical equilibrium and acid-base reactions

Reversible reactions and dynamic equilibrium

Not all chemical reactions 'go to completion', i.e. the reactants change completely to form products. Reactions do not only move in the forward direction, some reactions also move in the backward direction and products change back into reactants. These reactions are called reversible and are indicated by the ' $\rightleftharpoons$ ' sign.

Equilibrium is a term used to denote the balance between the forward and reverse reactions. At equilibrium there is no observable change; however the system is in constant motion. As fast as the reactants are converted into products, the products are converted back into reactants. No changes are apparent on a macro scale (e.g. concentrations of reactants and products are constant) but reactions continue on a molecular scale. This is called **dynamic equilibrium**.

Position of equilibrium

An equilibrium only applies as long as the system remains isolated. In an isolated system, no materials are being added or taken away and no external conditions are being altered.

The proportion of products to reactants in an equilibrium mixture is known as the position of equilibrium.

The position of equilibrium can be altered by changing:

- concentration of the reactants or products
- pressure in reactions involving gases
- temperature.

The effect of a change can be predicted using **Le Chatelier's principle**.

Effect of concentration

Consider the equilibrium:

$$2CrO_4^{2-}(aq) + 2H^+(aq) \rightleftharpoons Cr_2O_7^{2-}(aq) + H_2O(l)$$

yellow   orange

Adding more acid increases the concentration of H^+ ions, so the system will try and minimise this effect by decreasing the concentration of H^+ ions and the position of equilibrium will move to the right, forming more products (colour change from yellow to orange).

Grade boost

If a question asks what you would observe when an equilibrium is affected by a change in conditions, you will be expected to use information given and state any colour changes that would occur.

Effect of pressure and temperature

Consider the equilibrium:

$$N_2(g) + 3H_2(g) \rightleftharpoons 2NH_3(g) \qquad \Delta H = -92 \text{ kJ mol}^{-1}$$

In total, there are 4 moles of gas on the left-hand side and 2 moles of gas on the right-hand side. The side with the greater moles of gas is the side at the higher pressure.

If the total pressure is increased, the equilibrium will shift to minimise this increase. The pressure will decrease if the equilibrium system contains fewer gas molecules. Therefore the position of equilibrium moves to the right (4 moles to 2 moles) and increases the yield of ammonia.

If the temperature is increased, the system will try and minimise this increase. The system opposes the change by taking in heat, so the position of equilibrium moves in the endothermic direction. Since ΔH is negative, the forward reaction is exothermic and the backward reaction endothermic, so the equilibrium moves to the left decreasing the yield of ammonia.

Grade boost

If the concentration of a reactant is increased, the position of equilibrium moves to the right and more products are formed. Increasing the pressure moves the position of equilibrium to whichever side of the equation has fewer gas molecules. An increase in temperature moves the position of equilibrium in the endothermic direction.

Effect of catalyst

A catalyst does not affect the position of equilibrium, but equilibrium is reached faster.

quickfire

② Write down the formulae for the following:

  (a) sulfuric acid

  (b) ammonia

  (c) potassium hydroxide.

Grade boost

The higher the H⁺ ion concentration, the lower the pH and the stronger the acid.

≫ Pointer

You do not have to learn the definition of pH.

Acids and bases/alkalis

You should be familiar with many **acids**, e.g. hydrochloric acid, HCl; sulfuric acid, H_2SO_4; ethanoic acid, CH_3COOH. All these acids contain hydrogen. When an acid is added to water, the acid releases H⁺ ions into solution, e.g.

$$HCl(g) \rightarrow H^+(aq) + Cl^-(aq)$$

(In fact the H⁺ ion bonds with a water molecule to form the H_3O^+ ion.)

When aqueous **bases** or alkalis are added to water the hydroxide ion OH⁻ forms, e.g.

$$NaOH(s) \rightarrow Na^+(aq) + OH^-(aq)$$

$$NH_3(aq) + H_2O(l) \rightleftharpoons NH_4^+(aq) + OH^-(aq)$$

Note that an ammonia molecule accepts a proton from a water molecule, so the ammonia acts as a base and the water acts as an acid.

Acids react with alkalis (and bases) in neutralisation reactions. In solution the hydroxide ions from alkalis neutralise the protons from acids.

$$H^+(aq) + OH^-(aq) \rightarrow H_2O(l)$$

The pH scale

The acidity of a solution is a measure of the concentration of aqueous hydrogen ions, H⁺. However, these concentrations are very small and vary over a wide range (between 1 and 0.00000000000001 mol dm⁻³.)

The Danish chemist, Sorenson adopted a logarithmic scale to overcome this and called it the pH scale. He defined pH as:

$$pH = -\log_{10}[H^+] \quad \text{where } [H^+] \text{ is the concentration of H}^+ \text{ in mol dm}^{-3}.$$

The negative sign in the equation results in pH decreasing as the aqueous hydrogen ion concentration increases. If the H⁺ ion concentration is greater than 10^{-7} mol dm⁻³, the pH is less than 7.

Using the pH scale the acidity of any solution can be expressed as a simple more manageable number, ranging from 0 to 14. This is much more convenient for the general public when dealing with concepts of acidity.

Acidic solutions have pH values < 7

Basic solutions have pH values > 7

Neutral solutions have pH values of 7.

	Universal indicator colour							
	red	orange	yellow	green	green blue	blue	dark blue	purple
pH	0–2	3–4	5–6	7	8	9–10	11–12	13–14
	all these are acids			neutral	all these are alkalis			
	the stronger the acid the lower the pH				the stronger the alkali the higher the pH			

Carbon dioxide

Carbon dioxide is an acidic gas which reacts with water to form carbonic acid.

$$CO_2 + H_2O \rightleftharpoons H_2CO_3$$

Since rainwater contains carbon dioxide dissolved from the air it is acidic. If rainwater flows over or through ground containing limestone (calcium carbonate), the rainwater will react with the calcium carbonate and change some of it into soluble calcium hydrogencarbonate thus eroding the limestone.

$$H_2O(l) + CO_2(g) + CaCO_3(s) \rightarrow Ca(HCO_3)_2(aq)$$

When water containing calcium hydrogencarbonate evaporates, a deposit of calcium carbonate forms and this results in the formation of stalagmites and stalactites in limestone caverns.

$$Ca(HCO_3)_2(aq) \rightarrow H_2O(l) + CO_2(g) + CaCO_3(s)$$

» Pointer

There is a strong link between carbon dioxide and Topic 3 the role of Green Chemistry in helping to achieve sustainability. Carbon dioxide emissions are a major contributor to global warming.
The combustion of fossil fuels is the source of most of the carbon dioxide that is enhancing the greenhouse effect.

Sea water

About half of the carbon dioxide formed by burning fossil fuels dissolves in the oceans. The equilibrium may be written as:

$$H_2O(l) + CO_2(g) \rightleftharpoons H^+(aq) + HCO_3^-(aq) \qquad (1)$$

Another important equilibrium in the ocean is that between hydrogencarbonate and carbonate ions.

$$HCO_3^-(aq) \rightleftharpoons CO_3^{2-}(aq) + H^+(aq) \qquad (2)$$

Many animals in the ocean make shells of calcium carbonate using the equilibrium.

$$Ca^{2+}(aq) + CO_3^{2-}(aq) \rightleftharpoons CaCO_3(s) \qquad (3)$$

The pH of sea water remains fairly constant at a value between 7.5 and 8.5. This means that sea water is slightly alkaline. Its pH is maintained by the buffering action of dissolved carbon dioxide, hydrogencarbonate and carbonate ions and ions such as Ca^{2+} and Mg^{2+}.

There has been concern that the increase in carbon dioxide levels in the atmosphere from the combustion of fossil fuels may have a detrimental effect on the surface waters of the oceans.

Look at equilibrium 1 above. By Le Chatelier's principle, an increase in carbon dioxide will push the equilibrium to the right and increase the concentration of hydrogencarbonate ions and hydrogen ions. This will decrease the pH, i.e. make the water more acidic.

There are also fears that this increased acidity will lead to decreased calcification, i.e. production of shells and plates out of calcium carbonate, in marine organisms.

An increase in the concentration of hydrogen ions will affect equilibrium 2 moving it towards the hydrogencarbonate ion and decreasing the amount of carbonate ions present.

The reduction in carbonate ions will affect equilibrium 3 moving it to the left and the solid shell will tend to dissolve rather than form.

⊙≪≪≪ quickfire

③ By considering its interaction with water, explain how carbon dioxide can behave as an acid.

» Pointer

A buffer solution is one that resists a change in pH when contaminated with small amounts of acid or alkali. You do not have to learn this definition.

Acid-base titrations

An acid-base titration is a type of volumetric analysis where the volume of one solution, say, an acid, that reacts exactly with a known volume of another solution, say, a base, is measured. The precise point of neutralisation is measured using an indicator. One of these solutions must be a standard solution (i.e. one of which the exact concentration is known) or it must have been standardised.

In the analysis, you use the standard solution to find out information about the substance dissolved in the other solution. This information could be:

- the concentration of the solution
- its molar mass
- its formula
- the percentage of the substance in an everyday product.

Standard solution

A standard solution can only be made from a solid which is almost 100% pure, e.g. anhydrous sodium carbonate can be used but sodium hydroxide cannot as it reacts with atmospheric carbon dioxide.

A standard solution is prepared from a solid as follows:

- Accurately weigh the required mass of solid.
- Transfer *all* of the solid into a beaker, add water and stir until all the solid dissolves.
- Pour the solution carefully into a volumetric (graduated) flask and add water until just below the graduation mark.
- Add water drop by drop until the graduation mark is reached and mix the solution thoroughly.

Performing a titration

- Pour one solution, say, an acid, into a burette, making sure that the jet is filled. Read the burette.
- Use a pipette to add a measured volume of the other solution, say, a base, into a conical flask.
- Add a few drops of indicator to the solution in the flask.
- Run the acid from the burette to the solution in the conical flask, swirling the flask.
- Stop when the indicator just changes colour (this is the end-point of the titration).
- Read the burette again and subtract to find the volume of acid used (this is known as the titre).
- Repeat the titration, making sure that the acid is added drop by drop near the end point, until you have at least two readings that are within 0.20 cm^3 of each other and calculate a mean titre.

Preparing a standard solution

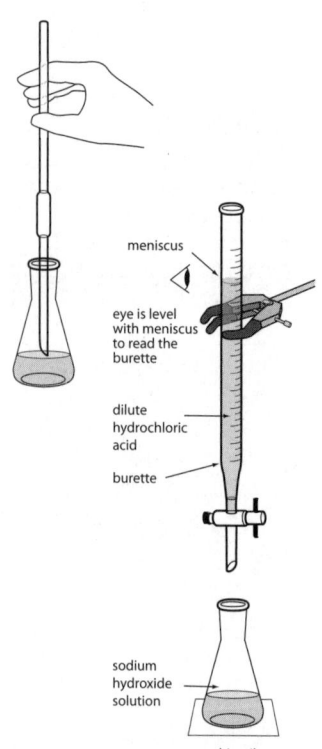

meniscus

eye is level with meniscus to read the burette

dilute hydrochloric acid

burette

sodium hydroxide solution

white tile

Performing a titration

Titration calculations

Example 1

20.0 cm³ of sulfuric acid was exactly neutralised by 24.0 cm³ of 0.950 mol dm⁻³ aqueous sodium hydroxide. Calculate the concentration of the acid.

$$H_2SO_4 + 2NaOH \rightarrow Na_2SO_4 + 2H_2O$$

(a) Calculate the amount, in moles, of NaOH that reacted

$n = c \times v = 0.95 \times 0.024 = 0.0228 (2.28.10^{-2})$

(Remember to divide 24 by 1000 to change it to dm³.)

(b) Use the equation to deduce the amount, in moles, of H_2SO_4 used

| from the equation | 2 moles NaOH | require | 1 mole H_2SO_4 |
| actual amounts used | 0.0228 moles NaOH | require | 0.0114 moles H_2SO_4 |

(c) Calculate the concentration, in mol dm⁻³, of H_2SO_4.

$$c = \frac{n}{v} = \frac{0.0114}{0.020} = 0.570 \text{ mol dm}^{-3}$$

Example 2

4.76 g of washing soda, which is hydrated sodium carbonate, were dissolved in water and the solution made up to 250 cm³. A 25.0 cm³ portion of this solution required 33.2 cm³ of 0.100 mol dm⁻³ aqueous hydrochloric acids solution for neutralisation. Calculate the percentage by mass of sodium carbonate in washing soda.

$$Na_2CO_3 + 2HCl \rightarrow 2NaCl + H_2O + CO_2$$

(a) Calculate the amount, in moles, of HCl that reacted

$n = c \times v = 0.1 \times 0.0332 = 0.00332 (3.32.10^{-3})$

(b) Use the equation to deduce the amount, in moles, of Na_2CO_3 used in the titration

| from the equation | 2 moles HCl | require | 1 mole Na_2CO_3 |
| actual amounts used | 0.00332 moles HCl | require | 0.00166 moles Na_2CO_3 |

(c) Calculate the amount, in moles, of Na_2CO_3 in the original 250 cm³ solution

25.0 cm³ of Na_2CO_3 contain 0.00166 moles

250 cm³ solution contains 10 × 0.00166 moles = 0.0166 moles

(d) Calculate the mass, in grams, of Na_2CO_3 in the original solution

$m = n \times M = 0.0166 \times 106 = 1.76 \text{ g}$

(e) Calculate the percentage by mass of Na_2CO_3 in the washing soda

$$\% = \frac{\text{mass } Na_2CO_3}{\text{mass washing soda}} \times 100 = \frac{1.76}{4.76} \times 100 = 37.0\%$$

≫ Pointer

In AS Chemistry, titration calculations should be structured similarly to the examples.

In A2 you may have to work out these steps for yourself.

Grade boost

Analysis of titration results follows a set pattern:
Calculate the moles of the solution for which the volume and concentration are given.

Use the mole ratio in the equation to find the moles of the other solution.
Change the moles into the answer required (concentration, molar mass, etc.)

 quickfire

④ 20.0 cm³ of aqueous sodium hydroxide was exactly neutralised by 16.0 cm³ of 0.250 mol dm⁻³ aqueous hydrochloric acid solution. Calculate the concentration of the sodium hydroxide.

$NaOH + HCl \rightarrow NaCl + H_2O$

Key Terms

Enthalpy, H, = the heat content of a system.

Enthalpy change, ΔH, = the heat added to a system at constant pressure.

The standard enthalpy change of formation, ΔH_f^θ, = the enthalpy change when one mole of a substance is formed from its constituent elements in their standard states under standard conditions.

The standard enthalpy change of combustion, ΔH_c^θ, = the enthalpy change when one mole of a substance is completely combusted in oxygen under standard conditions.

2.2 Energetics

Energy and enthalpy

Energy changes in chemical reactions are very important. The kind of life we lead depends on harnessing energy from different sources. There are many forms of energy. We shall be dealing with two: heat (a form of kinetic energy) and chemical energy (a form of potential energy).

In a chemical reaction existing bonds are broken and new bonds are made. This changes the chemical energy of atoms, and energy is exchanged between the chemical system and the surroundings as heat. To measure this energy the term **enthalpy, H**, is used.

Conservation of energy

The principle of conservation of energy states that energy cannot be created or destroyed, only changed from one form to another. Therefore if heat is released in a reaction, the amount of energy that leaves a system is exactly the same as the amount of energy that goes into the surroundings.

Enthalpy changes

Although enthalpy cannot be measured, an **enthalpy change**, ΔH, can easily be measured:

$$\Delta H = H_{products} - H_{reactants}$$

For a reaction that releases heat, i.e. heat is given out from the system to the surroundings, the enthalpy of the products is less than the enthalpy of the reactants so the sign of ΔH is negative. The reaction is said to be exothermic.

For a reaction that absorbs heat, i.e. heat is taken in from the surroundings to the system, the enthalpy of the products is more than the enthalpy of the reactants so the sign of ΔH is positive. The reaction is said to be endothermic.

Since enthalpy change for reactions depends on the conditions, for values to be compared, standard enthalpy changes measured under fixed conditions are used. Standard conditions are 298 K (25°C) and 1 atm (101 kPa).

Three very important enthalpy changes are:

- Standard enthalpy change of reaction, ΔH_r
- **Standard enthalpy change of formation, ΔH_f**
- **Standard enthalpy change of combustion, ΔH_c.**

You do not have to learn the definitions of these enthalpy changes but you will have to use them in calculating standard enthalpy changes.

If we are forming an element, such as $H_2(g)$, from the element $H_2(g)$, there is no chemical change. Therefore all elements in their standard state have a standard enthalpy change of formation of 0 kJ mol^{-1}.

 **Pointer**

The chemical system is the reactants and products. The surroundings are everything other than the system.

 **Grade boost**

Remember standard conditions are 298 K and 1 atm.

» Pointer

Standard state is the state of a substance under standard conditions.

Hess's Law

It is not always possible to measure the enthalpy change of a reaction directly. **Hess's Law**, which is based on the law of conservation of energy, gives a method for finding an enthalpy change indirectly.

The enthalpy cycle shows two routes for converting reactants to products. The first is a direct route and the second an indirect route via the formation of an intermediate.

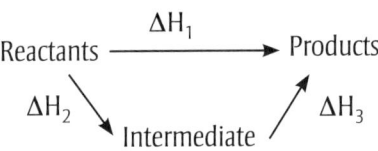

By Hess's Law the total enthalpy is independent of the route, so route 1 = route 2,

i.e. $\Delta H_1 = \Delta H_2 + \Delta H_3$

Example 1

Calculate the enthalpy of formation of ethane, C_2H_6, using the following information:

$$2C(s) + 3H_2(g) \rightarrow C_2H_6(g)$$

Substance	Enthalpy of combustion, ΔH_c^θ/kJ mol^{-1}
C(s)	−394
$H_2(g)$	−286
$C_2H_6(g)$	−1560

Construct an enthalpy cycle linking the reactants and products to the common combustion products

$$\begin{array}{ccc}
& \xrightarrow{\quad \Delta H \quad} & C_2H_6(g) \\
2C(s) + 3H_2(g) & & \\
(2 \times -394) \quad \text{Route 2} \downarrow & \text{Route 1} \downarrow & -1560 \\
+ & & \\
(3 \times -286) & \text{Combustion products } (CO_2 + H_2O) &
\end{array}$$

By Hess's Law route 1 = route 2

$\Delta H + (-1560) = [(2 \times -394) + (3 \times -286)]$

$\Delta H = [(2 \times -394) + (3 \times -286)] - (-1560) = -86$ kJ mol^{-1}

Or use the equation $\Delta H = \Sigma \Delta H_c(\text{reactants}) - \Sigma \Delta H_c(\text{products})$

$\Delta H = [(2 \times -394) + (3 \times -286)] - (-1560) = -86$ kJ mol^{-1}

Example 2

If the standard enthalpy change of formation of nitrogen oxide, NO, is 90.3 kJ mol^{-1} and that of nitrogen dioxide, NO_2, is 33 2 kJ mol^{-1}, calculate the enthalpy change for the reaction

$$2NO(g) + O_2(g) \rightarrow 2NO_2(g)$$

Use the equation $\Delta H = \Sigma \Delta H_f(\text{products}) - \Sigma \Delta H_f(\text{reactants})$

$\Delta H = (2 \times 33.2) - [(2 \times 90.3) + 0] = 66.4 - 180.6 = -114.2$ kJ mol^{-1}

Key Term

Hess's Law = states that the total enthalpy change for a reaction is independent of the route taken from the reactants to the products.

≫ Pointer

In an enthalpy cycle using ΔH_c, the direction of the arrows goes from the reactants and products to the common combustion products.

≫ Pointer

In an enthalpy cycle using ΔH_f, the direction of the arrows goes from the common elements to the reactants and products.

Grade boost

Using enthalpy changes of combustion
$\Delta H = \Sigma \Delta H_c(\text{reactants}) - \Sigma \Delta H_c(\text{products})$

Grade boost

Using enthalpy changes of formation
$\Delta H = \Sigma \Delta H_f(\text{products}) - \Sigma \Delta H_f(\text{reactants})$

quickfire

⑤ Given the following enthalpy changes of formation:

Substance	NH_3	O_2	NO	H_2O
ΔH_f^θ/kJ mol^{-1}	−46	0	90	−286

Calculate the enthalpy change for the reaction

$4NH_3(g) + 5O_2(g) \rightarrow 4NO(g) + 6H_2O(l)$

>> Pointer

Specific heat capacity, c, is the energy required to raise the temperature of 1 g of a substance by 1 K. (The value for water is 4.18 $Jg^{-1}K^{-1}$ and will always be given in a question.)

Measuring enthalpy changes

You cannot directly measure the heat content (enthalpy) of a system but you can measure the heat transferred to its surroundings. If a calorimeter is used the system is insulated thermally from its surroundings. The change in the temperature inside the calorimeter caused by the enthalpy change of the reaction can be measured with a thermometer.

If the temperature change is recorded and the mass and specific heat capacity of the contents of the calorimeter are known then the enthalpy change can be calculated.

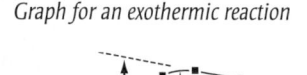

Calculating enthalpy changes

Enthalpy change can be calculated using the expression:

$$\Delta H = \frac{-mc\,\Delta T}{n}$$

m is the mass of the solution in the cup

c is the specific heat capacity of the solution

ΔT is the maximum temperature change

n is the amount, in moles, that has reacted.

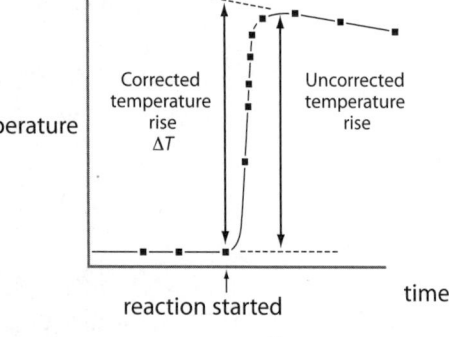

Graph for an exothermic reaction

temperature

Corrected temperature rise ΔT

Uncorrected temperature rise

reaction started   time

To obtain the maximum temperature change, allowances are made for heat lost (or gained) to (or from) the surroundings. Therefore, temperatures of the solution are taken for a short period before mixing and for some time after mixing. A graph of temperature against time is plotted and the maximum temperature is obtained by extrapolating the graph back to the mixing time.

It is important that the contents are well stirred during mixing to ensure that all the reactants react and that as rapid a reaction as possible takes place. If a solid is used it should be in powder form.

Since the mass of the solution is used in the expression to calculate ΔH it has to be measured accurately. The density of the solution can be assumed to be 1 g cm^{-3} therefore a burette or pipette can be used to measure the volume of the solution.

The mass of a solid is not added to the mass of the solution. However, if both reactants are solutions then the mass is equal to the total volume of the solutions.

Since the number of moles is needed to calculate the molar enthalpy change, the reactant that is not in excess has to be measured accurately. So the mass of a solid or the concentration of a solution must be known.

The coffee cup calorimeter

This is the simplest type of calorimeter and is suitable for changes that take place in aqueous solution, e.g. displacement reactions, neutralisations, dissolving a solid.

The expanded polystyrene insulates the solution inside the cup so the amount of heat lost or absorbed by the cup during the experiment is negligible. (It can be placed inside a beaker and lagged with cotton wool to improve insulation.)

Example

6 g of zinc was added to 25.0 cm³ of 1.00 mol dm⁻³ copper sulfate solution in a polystyrene cup. The temperature increased from 20.2°C to 70.8°C.

Calculate the enthalpy change for the reaction

$Zn(s) + CuSO_4(aq) \rightarrow ZnSO_4(aq) + Cu(s)$

Assume that the density of the solution is 1.00 g cm⁻³ and its specific heat capacity, c, is 4.18 Jg⁻¹K⁻¹.

Step 1 Calculate the enthalpy change for the amounts used in the experiment.

25 cm³ has a mass of 25 g

$\Delta T = 70.8 - 20.2$

$\Delta H = -mc\Delta T = -25 \times 4.18 \times 50.6 = -5288 \text{ J}$

Step 2 Calculate the amount, in moles, that reacted

$$\text{moles Zn} = \frac{m}{M} = \frac{6}{65.4} = 0.092$$

$$\text{moles CuSO}_4 = c \times \frac{v}{1000} = 1 \times 0.025 = 0.025$$

therefore $CuSO_4$ is not in excess and is used in the calculation

Step 3 Calculate the molar enthalpy change

$$\Delta H = \frac{-mc\Delta T}{n} = \frac{-5288}{0.025} = -211520 \text{ J mol}^{-1} = -211.5 \text{ kJ mol}^{-1}$$

This value is less than the actual value due to heat losses from the simple type of calorimeter used.

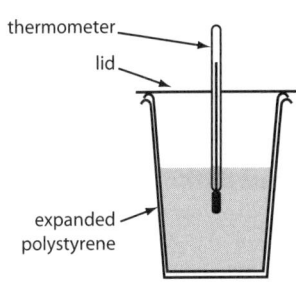

thermometer
lid
expanded polystyrene

>> *Pointer*

Remember the *m* in step 1 is the mass of the solution that is changing the temperature not the solid.

▲ **Grade boost**

Show all your working. Each step of the calculation will be worth a mark.

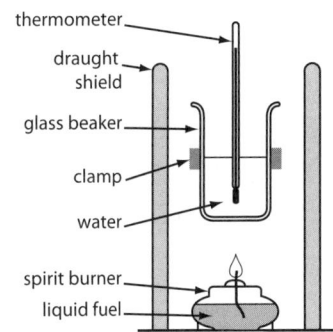

thermometer
draught shield
glass beaker
clamp
water
spirit burner
liquid fuel

Enthalpy change of combustion, ΔHc

The experimental determination of ΔH_c for a fuel can easily be carried out. A known mass of fuel is burnt in air to heat a known mass of water and the temperature change in the water is recorded.

There are two main sources of error in this experiment:

- Heat loss to the surroundings.
- Incomplete combustion of the fuel.

◉ ≪≪≪ quicKfire

⑥ The combustion of 1.50 g of ethanol raised the temperature of 500 cm³ of water by 19.5°C. Calculate the molar enthalpy change of combustion of ethanol, C_2H_5OH. (Assume c = 4.18JgK⁻¹)

Key Terms

Average bond enthalpy
= the average value of the enthalpy required to break a given type of covalent bond in the molecules of a gaseous species.

Bond enthalpy = the enthalpy required to break a covalent X–Y bond into X atoms and Y atoms, all in the gas phase.

Bond enthalpy

Bond enthalpy gives information about the strength of a covalent bond. Bond enthalpies show how much energy is needed to break different covalent bonds.

The H–H bond enthalpy is always the same because the H–H bond only exists in a H_2 molecule. However, C–H bonds exist in many different compounds. The actual value for the enthalpy change for a particular bond depends on the structure of the rest of the molecule, so the C–H bond strength varies across the different environment in which it is formed.

We therefore take average values for bond enthalpy derived from the full range of molecules that contain a particular bond.

Average bond enthalpies can be used to calculate standard enthalpy changes of reaction involving covalent compounds. The results of these calculations will not be as accurate as results derived from experiments with specific molecules. However, they usually give an accurate enough indication of the standard enthalpy change of reaction.

Example

Using average bond enthalpies, calculate the standard enthalpy change of reaction for the complete combustion of methane

$$CH_4(g) + 2O_2(g) \rightarrow CO_2(g) + 2H_2O(g)$$

Bond	C–H	O=O	C=O	O–H
Average bond enthalpy/kJ mol^{-1}	413	497	805	463

Step 1 Draw out each molecule

Step 2 Calculate the energy required to break the bonds (endothermic).

Bonds broken = 4 (C–H) + 2 (O = O) = (4 × 413) + (2 × 497) = 2646 kJ mol^{-1}

Step 3 Calculate the energy released when bonds are made (exothermic).

Bonds formed = 2 (C = O) + 4 (O–H) = (2 × 805) + (4 × 463) = 3462 kJ mol^{-1}

Step 4 Add together the energy changes

$\Delta H = \Sigma$(bonds broken) $- \Sigma$(bonds formed)

$\Delta H = 2646 - 3462 = -816$ kJ mol^{-1}

Grade boost

When asked to calculate enthalpy change using bond enthalpies, always draw out each molecule so that you can see the bonds broken and bonds made.

≫ Pointer

Breaking bonds requires energy, therefore is endothermic (so bond enthalpy is always positive). Making bonds releases energy so is exothermic.

2.3 Kinetics

Rates of reaction

Chemical kinetics investigates the rates at which chemical reactions take place.

For a reaction: $\text{rate} = \dfrac{\text{change in concentration}}{\text{time}}$ units: $\dfrac{\text{mol dm}^{-3}}{\text{s}} = \text{mol dm}^{-3}\text{ s}^{-1}$

If another variable, such as mass or volume is measured, the rate can be expressed in corresponding units such as g s^{-1} or $\text{cm}^3\text{ s}^{-1}$.

Usually for reactions:

- Rate is fastest at the start of a reaction since each reactant has its greatest concentration.
- Rate slows down as the reaction proceeds since the concentration of the reactants decreases.
- Rate becomes zero when the reaction stops, i.e. when one of the reactants has been used up.

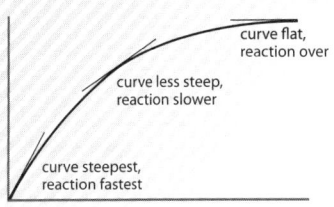
curve flat, reaction over
curve less steep, reaction slower
curve steepest, reaction fastest

Measuring rates of reaction

To measure the rate of a chemical reaction we need to find a physical or chemical quantity which varies with time. These are the main methods (all carried out at constant temperature).

- **Change in gas volume**   e.g. $Mg(s) + 2HCl(aq) \rightarrow MgCl_2(aq) + H_2(g)$

In a reaction in which gas is formed, the volume of the gas can be recorded using a gas syringe at various times.

- **Change in gas pressure**   e.g. $PCl_5(g) \rightarrow PCl_3(g) + Cl_2(g)$

Some reactions between gases involve a change in the number of moles of gas. The change in pressure (at constant volume) at various times can be followed using a manometer.

- **Change in mass**   e.g. $CaCO_3(s) + 2HCl(aq) \rightarrow CaCl_2(aq) + H_2O(l) + CO_2(g)$

- If a gas forms in a reaction and is allowed to escape, the change in mass at various times can be followed using weighing scales.

- **Change in colour**   e.g. $CH_3COCH_3(aq) + I_2(aq) \rightarrow CH_3COCH_2I(aq) + HI(aq)$

The intensity of the colour of the iodine can be monitored over time by using a colorimeter and hence its change in concentration can be measured.

- **Sampling**   e.g. $CH_3CO_2C_2H_5(aq) + H_2O(l) \rightarrow CH_3CO_2H(aq) + C_2H_5OH(aq)$

Samples of the reaction mixture are removed at various times. The reaction in each sample taken is slowed down significantly (quenched) by diluting in ice-cold water. Each sample is titrated against standard alkali and the concentration of ethanoic acid is calculated.

Calculating initial rates of reaction

We follow the rate of a reaction by measuring the concentration of a reactant (or product) over a period of time. The results obtained are plotted to give a graph. To find the initial rate, it is necessary to find the initial slope (gradient) of the line. At AS level the graph will always be a straight line to begin with, e.g.

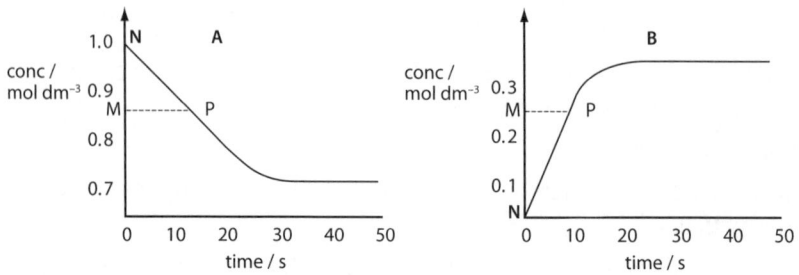

Grade boost

When drawing a graph, draw a line that best fits the points. All the points might not be on the 'best fit' line.

To find the gradient: at any convenient point, P, on the straight line draw a horizontal line MP to the y axis and draw a vertical line from M to the beginning of the slope, N.

For graph A:

$$\text{Rate} = \frac{\text{change in concentration}}{\text{time}} = \frac{\text{MN}}{\text{MP}} = \frac{(1 - 0.85)}{12} = \frac{0.15}{12} = 0.0125 \text{ mol dm}^{-3} \text{ s}^{-1}$$

For graph B:

$$\text{Rate} = \frac{\text{change in concentration}}{\text{time}} = \frac{\text{MN}}{\text{MP}} = \frac{0.25}{10} = 0.025 \text{ mol dm}^{-3} \text{ s}^{-1}$$

To find out the relationship between initial rate and the initial concentrations of the reactants, a series of experiments, in which the concentration of only one reactant is changed at a time, must be performed. The results of the initial concentrations and initial rates must then be compared.

The table below gives the experimental data for the reaction between propanone, CH_3COCH_3, and iodine, I_2, carried out in dilute hydrochloric acid.

$$CH_3COCH_3(aq) + I_2(aq) \rightarrow CH_3COCH_2I(aq) + HI(aq)$$

| Experiment | Initial concentrations / mol dm^{-3} | | | Initial rate / 10^{-4} mol dm^{-3} s^{-1} |
	I_2(aq)	CH_3COCH_3(aq)	HCl(aq)	
1	0.0005	0.4	1.0	0.6
2	0.0010	0.4	1.0	0.6
3	0.0010	0.8	1.0	1.2

Pointer

The amount of product formed in any reaction always depends on the amount of reactants even though the rate of reaction might not depend on the concentration of a particular reactant.

In experiments 1 and 2, only the concentration of iodine is changed and when it is doubled there is no change in the initial rate of reaction. Therefore the initial rate of reaction is independent of the initial concentration of aqueous iodine.

In experiments 2 and 3, only the concentration of propanone is changed and when it is doubled the initial rate of reaction also doubles.

Therefore the initial rate of reaction is directly proportional to the initial concentration of aqueous propanone.

Factors affecting reaction rates

- Temperature of a reaction
- Concentration of a solution (pressure of a gas)
- Surface area of a solid
- Catalyst
- Light (in some reactions e.g. $H_2 + Cl_2$, photosynthesis).

How these factors change the rate of a reaction can be explained using collision theory.

Collision theory

For a chemical reaction to take place, reacting molecules must collide. However, only a small fraction of the total number of collisions leads to a reaction. For a collision to be effective the molecules must collide in the correct orientation and have sufficient energy. The minimum energy needed is called the **activation energy**.

» Pointer

The reactant species that collide during chemical reactions may include molecules, atoms or ions. For the sake of simplicity we refer to these simply as 'molecules'.

Energy profiles

The activation energy may be shown on diagrams called energy profiles. These compare the enthalpy of the reactants with the enthalpy of the products.

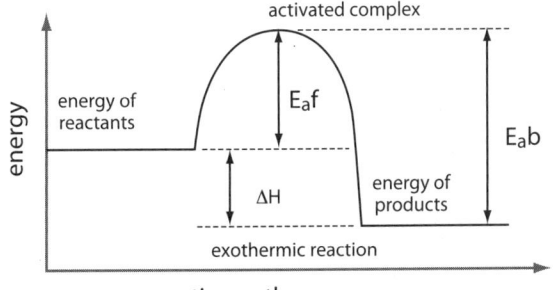

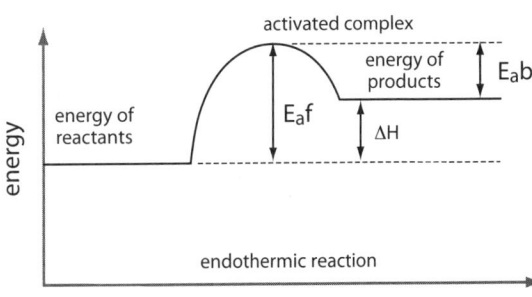

For an exothermic reaction, the reacting chemicals lose energy and heat is given out to the surroundings. Even though the products have a lower energy than the reactants, there still has to be an input of energy to break bonds and start the reaction.

For an endothermic reaction the enthalpy of the products is more than the enthalpy of the reactants and heat is taken in from the surroundings:

$$\Delta H = E_af - E_ab$$

Where E_af and E_ab are the activation energies of the forward and reverse reactions respectively.

For an exothermic reaction $E_af < E_ab$ and ΔH is negative.

For an endothermic reaction $E_af > E_ab$ and ΔH is positive.

» Pointer

The activated complex may be regarded as a transition state in which old bonds have partly broken and new bonds have partly formed.

⊙ ««« quicκϝιre

⑧ Draw an energy profile diagram for the following reaction and calculate the activation energy for the backward reaction.

$$H_2(g) + I_2(g) \leftrightarrows 2HI(g)$$
$$\Delta H = 53 \text{ kJ mol}^{-1}$$

E_a for the forward reaction = 183 kJ mol^{-1}

Grade boost

When explaining the effect of changing conditions on reaction rates always use the collision theory in your answer. Bullet points can be useful.

Effect of concentration (pressure) on reaction rates

If the concentration of a reactant increases, the reaction rate increases. There are more molecules in the same volume so there is a greater chance of the molecules colliding and therefore a greater chance that there will be more collisions with energy greater than the activation energy in a certain length of time.

For a gaseous reaction, increasing the pressure is the same as increasing the concentration.

For a solid increasing the surface area has the same effect.

 Grade boost

When explaining the effect of temperature on reaction rate you must include a reference to activation energy in your answer.

Effect of temperature on reaction rates

If the temperature of a reaction increases, the reaction rate increases. At higher temperatures the molecules have more kinetic energy and are moving faster. More molecules have an energy that is greater than the activation energy and more collisions take place in a certain length of time. This can be shown using energy distribution curves. In the diagram, temperature T_2 > temperature T_1.

 Grade boost

When drawing a distribution curve remember to draw the line representing activation energy far to the right on the energy axis and ensure that the distribution curve does not touch the axis.

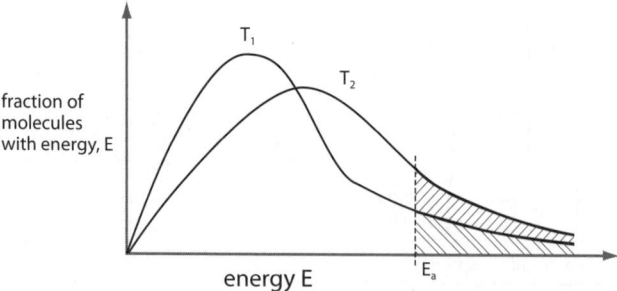

The areas under the two curves are equal and are proportional to the total number of molecules in the sample.

- The curves do not touch the energy axis.
- At the higher temperature, T_2, the peak moves to the right (higher energy) with a lower height.
- Only the molecules with an energy equal to or greater than the activation energy, E_a, are able to react.
- At the higher temperature, T_2, many more molecules have sufficient energy to react and so the rate increases significantly.

Catalysts

A catalyst increases the rate of a chemical reaction without being used up in the process. A catalyst does take part in the reaction but can be recovered unchanged at the end of the reaction.

A catalyst lowers the activation energy of the reaction by providing an alternative route for the reaction to follow. This can be shown on an energy profile diagram.

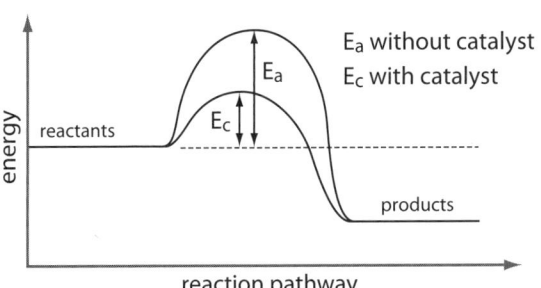

At the same temperature, a greater proportion of the reactant molecules will have sufficient energy to overcome the activation energy for a catalysed reaction. This can be shown on an energy distribution curve diagram.

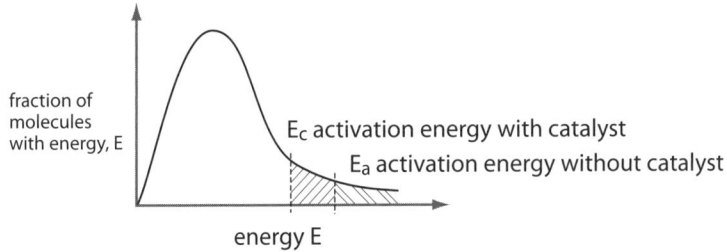

For a reversible reaction, a catalyst increases the rate of the forward and back reactions by the same amount, therefore it does not affect the position of equilibrium, but the position of equilibrium is reached quicker.

There are two classes of catalysts: heterogeneous and homogeneous.

Heterogeneous catalysts

A heterogeneous catalyst is in a different phase from the reactants. Many industrially heterogeneous catalysts are d-block transition metals. The transition metal provides a reaction site for the reaction to take place. Gases are adsorbed on to the metal surface and react and the products desorb from the surface. The larger the surface area, the better the catalyst works. Examples are:

- Iron in the Haber process for ammonia production.
- Vanadium(V) oxide in the Contact process within sulfuric acid manufacture.
- Nickel in the hydrogenation of unsaturated oils in the production of margarine.

Grade boost

A catalyst increases the rate of reaction by providing an alternative route of lower activation energy.

Pointer

A catalyst does not appear as a reactant in the overall equation of a reaction.

Grade boost

You should be able to state an example of a heterogeneous catalyst and a homogeneous catalyst in use.

quickfire

⑨ Give an example of a process that uses a heterogeneous catalyst, stating the process and naming the catalyst.

Key Term

An enzyme = a biological catalyst.

>> *Pointer*

Learn at least one example of homogeneous catalysts.

Grade boost

New developments in finding better catalysts are designed to:

- make the particular industry more profitable
- reduce energy and material costs
- make the chemical industry more green.

quickfire

⑩ Give two reasons why the use of enzymes in industrial processes reduces the effect on the environment.

>> *Pointer*

In industry, chemists try to make the highest possible yield of a desired product as quickly as possible. In many processes it may be uneconomical to keep the reaction going long enough for equilibrium to be reached, so a compromise between yield and rate is necessary.

Homogeneous catalysts

A homogeneous catalyst is in the same phase as the reactants. Homogeneous catalysts take an active part in a reaction rather than being an inactive spectator. Examples are:

- Concentrated sulfuric acid in the formation of an ester from a carboxylic acid and an alcohol.
- Aqueous iron(II) ions, Fe^{2+}(aq), in the oxidation of iodide ions, I^-(aq), by peroxodisulfate(VI) ions, $S_2O_8^{2-}$(aq).

Catalysts in industry

Most industrial processes involve catalysts. Industry relies on catalysts to reduce costs. A catalyst speeds up a process by lowering the activation energy of the reaction so less energy is required for the molecules to react, and this saves energy costs. Much of this energy is taken from electricity supplies or by burning fossil fuel so a catalyst also has benefits for the environment. If less fossil fuel is burnt, less carbon dioxide will be released during energy production.

A motor company has developed a new catalyst for petrol-powered cars that needs only half the precious metals of current catalysts. The new catalyst uses nano-technology, which prevents the metals from clustering. The Monsanto process of producing ethanoic acid from methanol originally used a pressure of 700 atm and a temperature of 300°C but by finding a new catalyst the process is now run at a pressure of 30 atm and a temperature of 150–200°C.

Enzymes

Enzymes are used in a variety of important industrial processes such as food and drink production and the manufacturing of detergents and cleaners. Some of the benefits are:

- Lower temperatures and pressures can be used, saving energy and costs.
- They operate in mild conditions and do not harm fabrics or food.
- They are biodegradable. Disposing of waste enzymes is no problem.
- They often allow reactions to take place which form pure products with no side reactions, removing the need for complex separation techniques.

One of the most extensive uses of enzymes is in biological washing powders. Since milder conditions can be used, less damage is caused to the clothes when washed. Because stains are made of different types of molecules, to break them down a range of enzymes are used in a washing powder.

Chemical equilibrium and reaction rate

Equilibrium tells us about the balance between reactants and products and how this balance is affected by changes in conditions. Kinetic data tells us how fast a reaction is and how it goes (mechanism of a reaction).

3a Application of the principles studied in Unit 1 to problems encountered in the production of chemicals and of energy

This topic requires you to apply the principles you have studied in Topics 1 and 2 to situations and problems encountered in the production of chemicals and energy. You will be supplied with data relevant to the situation, process or problem which you may not have met before and will be marked on your ability to analyse and evaluate the situation or problem, usually by answering a series of questions on the topic. Calculations may be required and clearly a basic understanding of equilibrium, energetics and kinetics will be important in tackling many of the questions.

There are no specific learning outcomes as such, what is needed is some practice and experience in applying the relevant sections of Topics 1 and 2 to the particular question. Here a study of recent questions in past CH1 papers may be useful:

Q8 in January 2010 dealt with the chemical removal of carbon dioxide from power station flues, the associated equilibria and temperature effects and a direct separation using a membrane formed by nanotechnology.

Q7 in May 2010 asked various questions about the Haber process and the formation and behaviour of ammonium fertilisers.

Q8 in January 2011 required an evaluation of the carbon dioxide concentration/global temperature relation.

>> *Pointer*

When tackling trial questions refer continually to the relevant outcomes in Topics 1 and 2.

Grade boost

The questions will be fairly straightforward but may need some flexibility and lateral thinking to see what is needed in what may be an unfamiliar problem.

quickfire

① Which three topics from Topics 1 and 2 will be useful when answering a question about the effects of changing temperature, pressure and catalyst on the yield and rate of production of polyethene?

3(b) and (c) The role of Green Chemistry and the impact of chemical processes

NB Under these topics you will usually be supplied with data and information as a basis for the discussion and evaluation in your answer.

The aim of Green Chemistry is to make the chemicals and products that we need with as little impact on the environment as possible. This means:

- Using as little energy as possible and getting this from renewable sources such as biomass, solar, wind and water rather than from finite fossil fuels such as oil, gas and coal.
- Using renewable raw materials such as plant-based compounds whenever possible.
- Using methods having high atom economy so that a high percentage of the mass of reactants ends up in the product giving little waste, see page 23 Atom economy.
- Developing better catalysts, such as enzymes, to carry out reactions at lower temperatures and pressures to save energy and avoiding specially strong plants, see page 40.
- Avoid using solvents, especially volatile organic solvents that are bad for the environment.
- Make products that are biodegradable at the end of their useful lives, where possible.
- Avoid using toxic materials if possible and ensure that no undesirable co-products or by-products are released into the environment.

Grade boost

Make it very clear when describing atom economy that this is based on **mass** or relative molecular mass and not on the number of molecules in the equation. Do not use the word 'amount', since this is connected via 'amount of substance' to the number of moles.

quickfire

② Define the term percentage atom economy.

Impacts

The chemical industry plays a major role in modern life as a producer of the materials we need, from semiconductors in mobile phones to fertilisers in agriculture. It is a large producer of economic wealth and a major employer. While occasional accidents get wide publicity, most operations are clean and safe, carefully controlled and regulated and located away from centres of population.

The major chemical reaction today is the combustion of fossil fuels in vehicles, homes and factories, not to produce a product but to provide energy. The unwanted carbon dioxide product is giving an accelerating increase in atmospheric levels, coupled to an increase in global warming. To minimise this is a major research area in chemistry and other sciences and an aim of Green Chemistry, as above.

Summary: Controlling and Using Chemical Changes

Atomic structure

Atomic number
The number of protons in the nucleus

Mass number
The number of protons and neutrons in the nucleus

Orbitals or sub-shells
- An orbital is a region of space in a fixed energy level where there is a high probability of finding an electron
- s sub-shells can hold 2 electrons
- p sub-shells can hold 6 electrons
- d sub-shells can hold 10 electrons

Ionisation energy (IE)
- The equation for the 1st IE is:
  $X(g) \rightarrow X^+(g) + e^-$
- IE generally increases across a period because of increased nuclear charge.
- IE decreases down a group because of increased shielding from inner electrons
- Successive IE measures energy needed to remove each electron in turn until all electrons are removed from an atom

Hydrogen spectrum
- Series of discrete lines which get closer as energy increases
- Balmer series formed by excited electrons emitting energy as they drop back down to n = 2 energy level
- IE can be calculated by measuring frequency of convergence limit of Lyman series (n = 1 energy level)

Radioactivity

Types
- α particle – cluster of 2 protons and 2 neutrons
- β particle – a fast-moving electron
- γ ray – high energy electromagnetic radiation

Half-life
Time taken for half of the atoms in a radioactive sample to decay

Mass spectrometer
Principles
- Vaporisation of sample before entering mass spectrometer
- Ionisation by bombarding with high energy electrons
- Acceleration – an electric field gets ions to correct speed
- Deflection – a magnetic field separates ions according to their mass/charge ratio
- Detection – ions pass through a slit and are detected by appropriate instruments

Uses
- Determination of relative abundance of isotopes
- Calculating relative atomic masses

Mass spectrum of chlorine
- Peaks at m/z 35 and 37, due to $^{35}Cl^+$ and $^{37}Cl^+$ respectively, in ratio of 3:1
- Peaks at m/z 70, 72 and 74, due to $(^{35}Cl-^{35}Cl)^+$, $(^{35}Cl-^{37}Cl)^+$ and $(^{37}Cl-^{37}Cl)^+$ respectively, in ratio of 9:6:1

Calculations

Relative atomic mass
Average mass of one atom of the element relative to one-twelfth the mass of one atom of carbon-12

Mole
Amount of any substance that contains the same number of particles as there are atoms in exactly 12 g of carbon-12

Empirical and molecular formulae
- Empirical formula shows the simplest whole number ratio of the amount of elements present
- Molecular formula is the actual formula of a particular compound

Equations involving moles
- For a solid;
  Number of moles (n) = $\dfrac{\text{mass of substance (m)}}{\text{molar mass (M)}}$
- For a solution;
  Number of moles (n) = concentration (c) $\times$ volume (v)
- For a gas;
  Number of moles (n) = $\dfrac{\text{volume of gas}}{\text{molar volume}}$
  (At 0°C molar gas volume = 22.4 dm^3)

Atom economy
$\dfrac{\text{mass of required product}}{\text{total mass of reactants}} \times 100$

Percentage yield
$\dfrac{\text{mass of product obtained}}{\text{maximum theoretical mass of product}} \times 100$

Equilibrium

- Dynamic equilibrium is when the forward and reverse reactions in a reversible reaction occur at the same rate
- Position of equilibrium refers to the proportion of products to reactants in an equilibrium mixture
- Le Chatelier's principle states that if a system at equilibrium is subjected to a change, the equilibrium tends to shift in order to minimise the effect of the change
- Carbon dioxide is an acidic gas. There are concerns that due to its interaction with water, increased carbon dioxide levels in the atmosphere may have a detrimental effect on the surface waters of the oceans and may cause decreased calcification in marine organisms

Acids and bases

- Acids are proton donors, bases are proton acceptors
- pH is a measure of acidity using manageable numbers with a scale from 0 to 14
- The lower the pH the higher the H^+ ion concentration and the stronger the acid
- When preparing a standard solution accurate weighing scales and a volumetric flask must be used
- During titrations, the key items of apparatus are burette, pipette and conical flask

Energetics

Enthalpy changes

- Enthalpy change, ΔH, is the heat added to a system at constant pressure
- If ΔH is negative the reaction is exothermic
  If ΔH is positive the reaction is endothermic
- Standard conditions are 298 K and 1 atm
- Hess's Law states that the total enthalpy change for a reaction is independent of the route taken from the reactants to the products
- Average bond enthalpy is the average value of the enthalpy required to break a given type of covalent bond in any molecule.
- Enthalpy changes of reaction involving covalent compounds are calculated using average bond enthalpies because the actual enthalpy change for a particular bond depends on the structure of the rest of the molecule

Measuring ΔH experimentally

- Use $\Delta H = \dfrac{-mc\,\Delta T}{n}$

  where ΔT, the corrected temperature change, can be found graphically by extrapolating back to the mixing time

  m is the mass of the solution not the solid

  n is the number of moles reacting, i.e. it is the number of moles of the reactant which is not in excess
- To prevent heat from escaping or being absorbed, a polystyrene cup with a lid is used instead of a beaker

Kinetics

- Rate of reaction is the change in concentration of a reactant or product per unit time
- Rates can be followed experimentally by a change in gas volume or pressure, a change in mass, a change in colour or by sampling
- For a chemical reaction to take place, molecules must collide successfully i.e. with a minimum amount of energy known as the activation energy
- An increase in concentration of a solution, pressure of a gas or surface area of a solid increases the rate of reaction because there is a greater chance of successful collisions in a certain length of time
- An increase in temperature increases the rate of reaction because more colliding molecules have the required activation energy

Catalysts

- A catalyst increases the rate of reaction by providing an alternative route of lower activation energy
- A heterogeneous catalyst is in a different phase from the reactants
  A homogenous catalyst is in the same phase as the reactants
- Enzymes are being increasingly used in industrial processes as they achieve some of the goals of green chemistry

CH2 Properties, Structure and Bonding

The usefulness of materials depends on their properties, which in turn depend on their internal structure and bonding. By understanding the relationship between these, chemists can design new useful materials. The types of forces between particles are studied along with several types of solid structures to show how these influence properties. The building blocks of materials are the elements and the relationship of their properties to their position in the Periodic Table is illustrated by a study of the elements of the s-block and Group 7.

An introduction to organic chemistry provides a way to understand how the properties of carbon compounds can be modified by the introduction of functional groups. A further topic deals with analytical techniques that use mass spectral data and characteristic infrared frequencies in the elucidation of structure.

Revision checklist

Tick column 1 when you have completed brief revision notes.
Tick column 2 when you think you have a good grasp of the topic.
Tick column 3 during final revision when you feel you have mastery of the topic.

4 Bonding

4.1 Chemical bonding

Bonds may be represented by 'dot and cross' diagrams in which the outer electrons of one atom forming the bond are shown as dots or open circles and those of the other atom by crosses.

Covalent bonding – each atom gives one   H $\overset{\circ}{\times}$ H   electron to the (single) bond pair having opposed spins.

Coordinate bonding – a covalent bond electrons come from the same atom.

Ionic bonding – one atom gives one or to the other and the resulting cation and one another electrically. Usually there together in a solid lattice.

in which both

more electrons
anion attract
are many ions

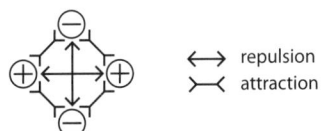

Attractive and repulsive forces

All bonding results from electrical attractions and repulsions between the protons and electrons, with attractions outweighing repulsions.

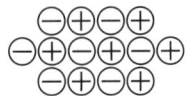

⟷ repulsion
⟩⟨ attraction

In covalent bonds the electrons in the pair between the atoms repel one another but this is overcome by their attractions to BOTH nuclei. If atoms get too close together the nuclei and their inner electrons will repel those of the other atom so that the bond has a certain length. Also the electron spins must be opposite for the bonds to form.

In ionic bonding cations and anions are arranged so that each cation is surrounded by several anions and vice versa to maximise attraction and minimise repulsion. Again repulsions from inner electrons and nuclei prevent the ions from getting too close together.

Key Terms

Covalent bond = has a pair of electrons of opposed spin shared between two atoms.

Coordinate bond = a covalent bond where both electrons come from one of the atoms.

Ionic bond = formed by the attractions between positive ions (cations) and negative ions (anions).

Grade boost

Be very careful when drawing 'dot and cross' diagrams to include all electrons, distinguish between electrons from the different atoms and include any charges. You may be asked to omit any inner electrons.

quickfire

① Draw one simple example of each of the three types of bond.

Grade boost

As you will see from the Quickfire question, it is the *difference* between the electronegativity values that is important and not their actual values.

 Pointer

Note that partial charges are shown by using the Greek lower case delta δ.

quickfire

② Using the electronegativity values given, arrange the bonds below in order of INCREASING polarity.

Values: Mg 1.2, H 2.1, Br 2.8, O 3.5, F 4.0

Bonds: Br–F, H–Br, Mg–O, O–H, F–F

Electronegativity and bond polarity

In a covalent bond the electron pair is not usually shared exactly evenly between the two atoms unless they are the same. Thus one atom will take up a slightly negative charge, the other becoming slightly positive and the bond is now said to be polar. These small charges are written over the atoms using the symbols $\delta+$ and $\delta-$ as shown below:

$$\overset{\delta+}{H} \text{------} \overset{\delta-}{F}$$

Coordinate bonds are always polar, since the atom giving both electrons to the bond cannot completely lose its rights over one electron.

Bond polarity is governed by the difference in **electronegativity** between the two atoms forming the bond. Electronegativity is a measure of the ability of an atom in a covalent bond to attract the electron pair and, on one scale, ranges from 0.7 in Cs to 4.0 in F, with Cs being electropositive and F highly electronegative.

Thus almost all bonds joining atoms that are not identical will be polar to some extent and, at the other extreme, most ionic bonds have some covalent character, but it is simpler for us to treat ionic bonds as completely ionic and take covalent bonds as having varying degrees of polarity.

The following diagrams show the electron density distribution for some different bonds.

Bond	NaF	H–Cl	H–H
Electronegativity difference	3.1	0.9	0.0
Electron distribution	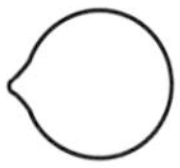	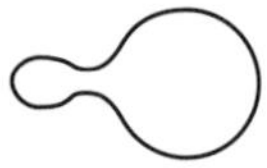	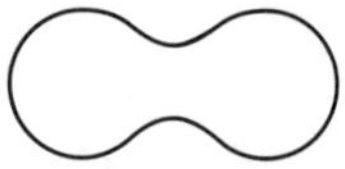

The bond in hydrogen chloride, for example, is about 19% ionic.

4.2 Forces between molecules

It is most important to distinguish bonding BETWEEN molecules – INTERMOLECULAR bonding – and bonding WITHIN molecules – INTRAMOLECULAR bonding.

Intermolecular bonding is weak and governs physical properties such as boiling temperature; bonding within molecules is strong and governs chemical reactivity. In methane, for example, the forces between the molecules are very weak and the molecules separate, i.e. the liquid boils at –162 degrees, but the C–H bonds are very strong and need a temperature of around 600 degrees before they will break.

Intermolecular bonding is caused by electrical attraction between opposite charges. Although the molecule may be neutral overall it contains positive and negative charges (electrons and protons) and if the electronegativities of the atoms in the molecule (see Topic 4.1) are not the same, the molecule will have a **dipole** with parts that are relatively positive and negative in charge. If these dipoles arrange themselves so that the negative region of one molecule is close to the positive region of another molecule, there will be a net attraction between them.

$$\delta+ \text{———} \delta- \quad \delta+ \text{———} \delta-$$
$$\delta- \text{———} \delta+ \quad \delta- \text{———} \delta+$$

δ is a permanent partial charge

Even molecules with no dipole show intermolecular bonding, e.g. helium atoms come together to form a liquid at 4K. This is because the electrons are in constant motion around the nuclei so that the centres of positive and negative charge do not always coincide and give a fluctuating dipole. These come into step with one another as one dipole induces an opposite dipole in a nearby molecule giving an attraction between them.

$$\delta\delta+ \text{———} \delta\delta- \quad \delta\delta+ \text{———} \delta\delta-$$
$$\delta\delta- \text{———} \delta\delta+ \quad \delta\delta- \text{———} \delta\delta+$$

$\delta\delta$ is a fluctuating induced charge

To sum up we have here two types of intermolecular bonding, the first dipole–dipole and the second induced dipole–induced dipole and these two together are called **van der Waals forces**.

Strength

Bonding inside molecules is some 100 times stronger than between them with van der Waals strengths being around 3kJ per mol.

Key Terms

Dipole = separation of change within a molecule. Electrical charges are not balanced so that one part has a partial negative charge and another an equal positive charge.

van der Waals forces = weak intermolecular forces made up of dipole–dipole and induced dipole–induced dipole forces of attraction.

Grade boost

Careful and accurate diagrams are a good way of getting marks in this section and the next one on hydrogen bonding.

quickfire

③ Explain why liquid nitrogen boils at 77K, although the bond between nitrogen atoms in the molecule is very strong.

Key Term

Hyrdrogen bond = The hydrogen bond is a relatively strong intermolecular bond having a hydrogen atom joined to a very electronegative element in a molecule and bonding to another electronegative element in another molecule.

⚑ Grade boost

1. Draw a very careful diagram of, e.g. O–H---N, showing the delta plus and delta minuses and the longer dotted H bond.

2. Make it very clear that a hydrogen bond is only a RELATIVELY strong bond compared with a van der Waals bond.

Hydrogen bonding

This is a special intermolecular bonding force that only occurs between molecules that contain hydrogen atoms bonded to very electronegative elements having lone pairs, such as fluorine, oxygen and nitrogen. Although weak compared to bonding inside molecules, it is much stronger than van der Waals forces. Typical strengths for **hydrogen bonds** would be 30 kJ per mol as against 3kJ for van der Waals and 300kJ for bonding within molecules.

Hydrogen bonding is stronger than that of van der Waals since the small hydrogen atom is sandwiched between two electronegative elements and allows close approach.

We see that the hydrogen atom is especially $\delta+$ being attached to the electronegative oxygen atom so that the oxygen atom on the other molecule is attracted closely to it. Also the bonding is strongest when the three atoms are in a straight line, and note that the internal O–H bond in the molecule is shorter than the dotted hydrogen bond connecting to the other molecule. Since oxygen has two lone pairs and two hydrogen atoms, a tetrahedral hydrogen-bonded structure is formed.

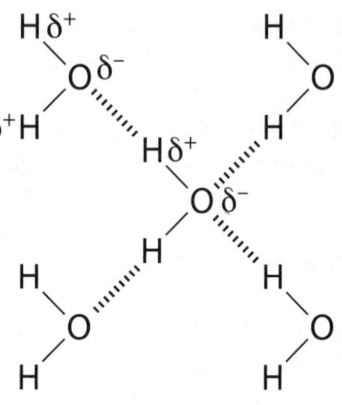

Hydrogen bonding in water

Effects of hydrogen bonding on boiling temperature and solubility

Melting and, especially, boiling temperatures increase with the strength of intermolecular forces. With van der Waals forces there is a steady increase with molecular mass and also as dipoles become larger.

The boiling temperature diagram given shows that the stronger hydrogen bonding completely bucks the trend.

In water the molecules freely hydrogen bond with neighbours, these bonds must be largely broken before boiling can occur and thus more energy, i.e. a higher temperature, is needed. There are many similar examples.

◉≪≪≪ quickfire

④ Arrange the following types of bonding in order of increasing strength: covalent bond inside a molecule, van der Waals bond, hydrogen bond.

⑤ Draw a possible hydrogen-bonded structure for liquid ammonia.

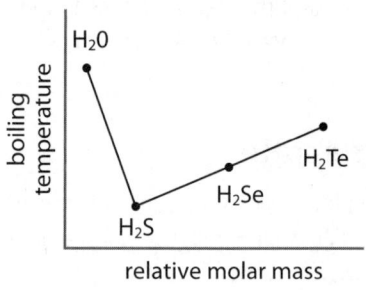

4.3 Shapes of molecules

The shapes of covalent molecules with more than two atoms and their ions are governed by the electron pairs around the central atom such as C in CH_4. These may be **bonding pairs** holding the atoms together in covalent bonds or **lone pairs** on the central atom that are not usually involved in covalent bonding.

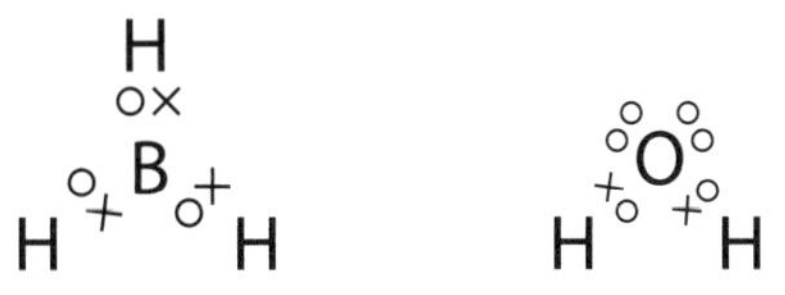

120°   104.5 °

Since all electron pairs repel one another, the molecular shape taken up is that allowing the pairs to keep as far away from each other as possible to minimise the repulsion energy. While bonding pairs are spread out between the two atoms bonded in the molecule, lone pairs stay close to the central atom and so repel more than bonding pairs, giving the repulsion sequence:

lone pair – lone pair > lone pair – bond pair > bond pair – bond pair.

Thus in NH_3 with one lone pair and three bonding pairs the repulsion between the lone pair and the bond pairs is greater than that between the bond pairs themselves so that the H–N–H angle is closed up from 109° to 107°.

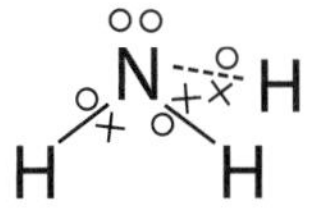

107°

These ideas are applied in the Valence Shell Electron Pair Repulsion (VSEPR) Theory that follows.

Key Terms

Bond pair = two electrons having opposed spins that bond two atoms in a molecule together by a covalent or coordinate bond.

Lone pair = two electrons having opposed spins that belong to one atom only and are not involved in bonding to another atom.

quickfire

⑥ Insert and label the bond pairs and lone pairs in the molecule below.

VSEPR theory

Valence Shell Electron Pair Repulsion (VSEPR) theory lets us predict the shape of simple molecules in which bonded atoms are arranged around a central atom. The valence shell is the electron shell in which bonding occurs. In VSEPR the number of electron pairs is first found to give the general shape of the molecule, since the repelling pairs keep as far away from one another in space as possible. It is worth repeating the repulsion sequence on the shapes of molecules from page 50:

lone pair – lone pair > lone pair – bond pair > bond pair – bond pair

and seeing how it applies in the figure on page 51.

The shape follows directly from the number of pairs as below:

No of pairs	Shape	Bond angle	Example
2	linear	180°	$BeCl_2$
3	trigonal planar	120°	BF_3
4	tetrahedral	109.5°	CH_4
5	trigonal bipyramid	90°/120°	PCl_5
	square planar	90°	
6	octahedral	90°	SF_6

Pointer

Learn and understand the shapes for each number of bonding pairs.

Grade boost

Count the electron pairs carefully and remember that it is only the pairs around the central atom that affect the shape, so that in CF_4, for example, all the lone pairs on the fluorine atoms are ignored, leaving just the four bonding pairs and a tetrahedral shape.

Secondly, the exact angles between the bonds will change somewhat depending on the repulsion sequence above. Thus in water with two lone pairs the normal tetrahedral bond angle of around 109° for H–O–H is repelled down to 104° by lone pair – bond pair repulsion.

You should be able to predict the shape of any simple molecule given its formula using VSEPR. Note that the same rules apply where the covalent molecule is an ion such as NH_4^+ where all electrons are now in bond pairs and the H–N–H bond angles are all 109.5°.

Note also that you are asked:

1) to know and explain the shapes of BF_3, CH_4, NH_4^+ and SF_6, and
2) predict and explain the shapes of other simple species having up to six electron pairs in the valence shell of the central atom.

quickfire

⑦ Predict the general shapes of BeF_2, BH_4^-, CH_3Cl and PF_5.

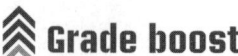

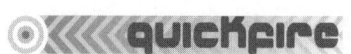

4.4 Solubility of compounds in water

A substance B will be soluble in C if the attractions between B and C molecules are greater than those between B and B and between C and C molecules.

If B is a hydrocarbon and C is water this will not be true and 'oil and water do not mix' but form separate layers. The water molecules attract one another through hydrogen bonding [see above] and hydrocarbon molecules interact more weakly through van der Waals forces. However, in alcohols O–H bonds have been added to the hydrocarbon structure so that hydrogen bonding now occurs between the alcohol and water and the alcohol dissolves in the water provided that its hydrocarbon chain is not too long.

Thus ethanol and the smaller alcohols dissolve in water up to a carbon chain length of about four atoms. Water dissolves many inorganic salts such as NaCl due to a strong interaction between the ions in the salt and the water dipoles.

The **polar** water has δ^- oxygen atoms and δ^+ hydrogens, the oxygen regions of several water molecules align themselves around cations and the hydrogens around the anions. The attractions of all these are sufficient to pull the ions from the solid lattice so that the salt dissolves in water.

Solubility terms

Solubility may be expressed as concentrations in either grams/dm^3 or mols/dm^3 and solubility in mol dm^{-3} = solubility in g dm^{-3} divided by the molar mass.

A **saturated solution** is one that has the maximum possible concentration of **solute** at the given conditions. Knowing the actual concentration is an important part of all chemical procedures.

Soluble salts may be recovered from aqueous solution by, e.g., evaporating water from a warm solution until it is saturated and then cooling to allow the salt to crystallise out.

5 Solid structures

Ionic, covalent and metallic

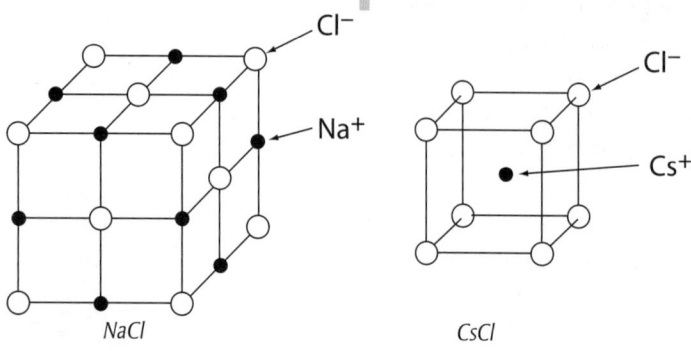

NaCl

CsCl

NB. Anions are larger than cations so only the cation size controls the crystal co-ordination number.

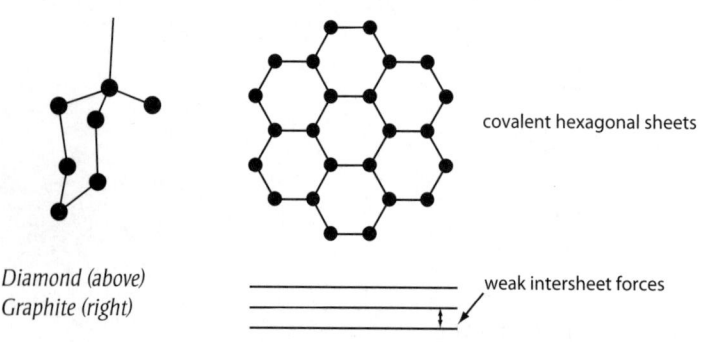

covalent hexagonal sheets

Diamond (above)
Graphite (right)

weak intersheet forces

You should *know* and be able to describe the ionic crystal structures of NaCl and CsCl, the covalent structures of diamond and graphite and know that iodine forms a molecular crystal but not know its structure.

In ionic halides, oppositely charged ions pack around one another in such a way as to increase the bonding energy by maximising electrostatic attraction and minimising repulsion. Thus each cation is surrounded by 6 or 8 anions and vice versa, the actual number depending only on the relative sizes in the chlorides. Eight chloride anions can fit around the larger Cs^+ ion while the smaller Na^+ can only accommodate six. The **crystal co-ordination numbers** are therefore 8:8 in CsCl and 6:6 in NaCl.

With diamond, the tetrahedral strong covalent arrangement and build up of a three-dimensional giant structure should be shown while, with graphite, layers made up of covalent hexagons held together by weak forces are needed.

In solid iodine it is very important to distinguish between the strong covalent bonds holding iodine atoms together in the I_2 molecule and the weak intermolecular forces that hold the I_2 units in the molecular crystal.

Grade boost

While artistic skill is not important in drawing the structures, the charges on all ions must be shown and the crystal co-ordination number must be evident in the diagrams.

Grade boost

A common source of error is not to make it clear that there are two types of bonding in solid iodine, the strongish covalent bond holding the I atoms in I_2 and the weak van der Waals forces holding the I_2 units in the crystal as above.

Metals

While the actual structures of metals are not required, you need to understand the general concept that atoms of metallic elements each donate one or more electrons to form a delocalised electron sea or gas that surrounds the packed positive ions so formed and binds them together through the attraction between opposite charges.

A sketch of bonding in metals

Nanoparticles, carbon nanotubes and smart materials

Nanoparticles

These may be defined as particles having one or more dimensions of about 100 nanometres (nm) or less. They thus lie in size between ordinary fine particles at around micron (10^{-6}m) size and molecules at around 1nm.

fine particle   nanoparticle   molecule

In some nanoparticles, such as carbon nanotubes, one dimension may be much larger, e.g. 2 mm while the tube diameter is 2nm. Below 100 nm the nanoparticles have properties that are different from the bulk material. For example, a nanoparticle of normally yellow gold is red in colour and melts at a temperature 700° lower than the bulk material. Many uses are being found for nanoparticles in medicine, electronics, biotechnology, clothing, etc.

Carbon nanotubes

These are tubes having a diameter of around 1 nanometre and a length millions of times greater. They are made of one atom layer graphite structures, called graphene, rolled into a seamless cylinder.

The tubes may be metallic or semiconductors, are very strong, hard and stiff, very good conductors of heat and very good conductors of electricity along the tubes when metallic. They are used as composites in polymers in engineering, including bio-engineering, and clothing, in electronic circuits, batteries and fuel and solar cells, etc.

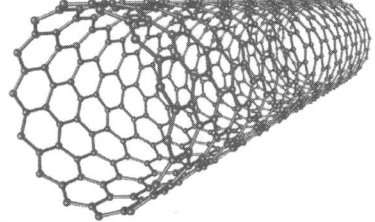

⚑ Grade boost

You are not expected to have any detailed knowledge of the structures of carbon nanotubes but up-to-date examples of new properties and uses found in the literature or online will help to impress the examiners. Similarly, good examples of smart materials could be valuable.

quickfire

① Suggest why carbon nanotubes may be good electrical conductors in one dimension.

Smart materials

These have properties that can be changed in a controlled way by a change in conditions such as temperature, pH, stress or electric or magnetic fields.

Examples include shape memory alloys and polymers that change shape reversibly under changes in temperature, stress or magnetic fields. Also photochromic or thermochromic materials that change colour, as in sunglasses that darken in sunlight. Voltage changes cause changes of shape in piezoelectric materials (and vice versa) and changes of colour in liquid crystal displays. There are also self-healing materials that repair damage caused by wear and tear.

⚑ Grade boost

Get clear in your mind how the properties of solids relate to their molecular structures, and have some examples ready.

Structure and physical properties

It is important to be able to explain the properties of all the solid types above in terms of their structures.

The giant ionic structures such as the chlorides are generally hard, brittle and high-melting due to the strong ionic bonds. There is no electrical conduction in the solid state since the ions are fixed in the crystal but the molten salts and aqueous solutions of them do conduct since they are now free to move when a voltage is applied. Ionic solids may or may not be soluble in water, depending on energetic or chemical reaction factors, but most ionic chlorides are soluble.

The covalent giants, diamond and graphite, are very high-melting and insoluble in water; diamond is very hard, with each carbon atom being covalently bonded to four others and forming a three-dimensional structure in space, but the weak layer structure in graphite renders it softer and useful as a lubricant. Also graphite conducts electricity owing to the π electron delocalisation in the ring plane while diamond and iodine do not. Iodine is soft and volatile since the I_2 units are held together only by weak van der Waals forces.

In carbon nanotubes the rolled graphite layer structure makes them the strongest and stiffest materials known. As with graphite, their properties are not usually the same in different directions in space owing to the nature of the bonding.

Electron delocalisation in metals gives good electrical and thermal conductivity but their melting temperatures and hardness increase with the number of electrons per atom involved in bonding, e.g.

Metal	Na	Ca	V
no of bonding electrons	1	2	5
melting temp. °C	98	850	1900

Sodium is a soft metal that is easily cut with a knife but vanadium is very hard. Also remember mercury, melting temperature minus 39°C!

In practice the actual properties of a solid depend not only on the bonding at the atomic level but on the way in which the units, such as the carbon nanotubes, are held together.

6.1 The Periodic Table

Basic structure

An understanding of the general trends in properties and behaviour gives us great predictive power. The chemistry of the elements is governed largely by their outer electrons so that arranging elements in groups according to their outer structure simplifies study of their behaviour.

Ionisation energy (IE) and electronegativity (x) increase diagonally across the Table (i.e. across a period or down a group), e.g. IEs for Cs and F are 376 and 1680 kJ.

Electrons are thus readily lost in the s block giving cations in ionic compounds; entering the p block in group 3, IEs become too high so that electron sharing (covalency) is usual, but the more electronegative elements of groups 6 and 7 can accept electrons to form anions in ionic compounds.

Valency normally rises with group number to a maximum of four and then falls (8 minus the group number) to one in group 7.

Elements are generally metals when IEs are low in the left and lower regions of the Table and the d block, transition, elements and non-metals in the high IE, upper right portion: semiconductor elements, e.g. Si, are found between these two regions.

Melting temperature trends are more complex, depending on atomic mass, type of solid structure and bond type but decrease down group 1, rise down group 7 and increase across a period up to group 4 (C > 3500°C) and then drop sharply as elements form diatomic molecules held by weak intermolecular forces.

>> *Pointer*

See also Topic 1 sections on ionisation energies, electronic structures and orbital shapes.

⊙ꗇꗇꗇ **quickfire**

① Two of the ions below are not met in ordinary chemistry. Identify these two and state why they are not seen.
Ca^{2+} B^{3+} C^{4+}

⊙ꗇꗇꗇ **quickfire**

② Explain why IEs increase across a period but decrease down a group.

	s block		d block	p block					
group	1	2	transition metals	3	4	5	6	7	8
ox. no.	+1	+2					−2	−1	inert gas
redox	reducing						oxidising		
ions	cations						anions		
oxides	basic						acidic		
m.t.	m.t. dec.	$M(OH)_2$ sol. inc. MSO_4 sol. dec	m.t. inc. to gp 4				m.t. inc.		
			METALS	NON-METALS					

 IE and x increase

Trends in the Periodic Table

Redox

Many chemical reactions involve the loss or gain of electrons, a species being **oxidised** if it loses electrons and **reduced** if it gains them. Since electrons do not vanish or appear from nowhere, all these reactions involve a transfer of electrons from the species being oxidised to the one being reduced, e.g. in $Na + \frac{1}{2}Cl_2 = Na+ + Cl^-$ the Na is oxidised losing an electron and the Cl is reduced gaining an electron.

The popular mnemonic OILRIG is helpful if used carefully as with the atom speaking 'oxidised I lose electrons, reduced I gain electrons'. Confusion is very common.

Oxidation numbers (states)

This is a useful accounting system for **redox** with simple rules:

1. All elements have an oxidation number of zero.
2. Hydrogen in compounds is usually +1 or I.
3. Oxygen is usually −2 or -II.
4. Group 1 and 2 elements in compounds are I and II respectively.
5. Group 6 and 7 elements in compounds are usually -II and -I respectively.
6. An element bonded to itself is still 0.
7. The oxidation numbers of the elements in a compound or ion must add up to zero or the charge on the ion.

IMPORTANT

The oxidation number does not imply a charge, e.g. in MnO_4^- the oxidation numbers are Mn (+7) O_4 (−2 x 4) giving an overall charge of minus 1. The Mn is **not** 7+.

Watch out for the common error described in the Grade Boost.

Key Terms

A reducing agent = gives an electron to another species and is therefore oxidised by its loss.

An oxidising agent = removes an electron from another species and is therefore reduced.

Redox = a chemical reaction in which an electron is transferred from one species – the reducing agent – to another species, which is reduced by receiving the electron.

Grade boost

Note that the words redox and oxidation have not necessarily anything to do with oxygen but only with electron transfer.

③ a) Write the oxidation numbers for the four species in the reaction $Na + \frac{1}{2}Cl_2 = Na^+ + Cl^-$.

b) Which one of the following correctly gives the oxidation state of the chloride ion?
I, 1-, -I, 0.

c) Assign oxidation numbers to the elements in the following compounds: Na_2SO_4, NF_5, O_3, C_2H_6.

≫ Pointer

Redox studies will be carried out in more detail in the A2 year.

6.2 Trends in properties of s-block elements Groups 1 and 2

s-block elements

The elements are all reactive electropositive (low electronegativity) metals forming cations with oxidation numbers 1 or 2 respectively.

Oxides are formed with oxygen/air as in $Ca + \frac{1}{2}O_2 \rightarrow CaO$.

Hydrogen is liberated with water and an oxide or hydroxide formed;
$$Na + H_2O \rightarrow NaOH + \tfrac{1}{2}H_2.$$

The reaction of group 2 elements with acids is similar except that a salt is formed as in $Mg + 2HCl \rightarrow MgCl_2 + H_2$ and the elements in both groups show their typical reaction as reducing agents, donating electron(s) to reduce the acid or water to hydrogen and being themselves oxidised.
$$Mg(0) + 2H^+(+1)Cl^-(-1) \rightarrow Mg^{2+}(+2) + H_2(0) + 2Cl^-(-1)$$

In all these cases reactivity increases down the group and group 1 elements are more reactive than group 2.

The oxides are all basic, i.e. they react with acids to give salts, as in $CaO + 2HCl \rightarrow CaCl_2 + H_2O$.

Remember that the group 2 hydroxide formulae are $M(OH)_2$ since (OH) is -1, i.e $[O(-2)H(+1)]^{-1}$

While group 1 salts are all soluble the reactions of group 2 ions with OH^-, CO_3^{2-} and SO_4^{2-} give a variety of results that must be known. $Mg(OH)_2$ is insoluble in water but solubility increases *down* the group; $BaSO_4$ is insoluble and solubility increase *up* the group; all the carbonates are insoluble.

Flame colours: All of the common elements of groups 1 and 2 except Mg show characteristic flame colours that must be *known* and that are useful in qualitative analysis.

You should be aware of the great importance of calcium carbonate in both living and inorganic systems and of calcium phosphate minerals in living bones and skeletons. Calcium and magnesium ions play a vital role in the biochemistry of living systems – chlorophyll, muscle operation, etc., and the carbonates exist in huge amount in rocks – chalk, limestone, dolomite.

>> **Pointer**

A thorough understanding of the concepts and trends in 6.1 will make the work here much easier.

Grade boost

Be careful to write correct formulae for group 1 and 2 compounds, e.g. KOH, $Mg(OH)_2$, Na_2SO_4, $CaCO_3$ – mistakes look bad.

quickfire

④ i) Match the colours given with the correct element.

Colour: yellow, brick red, apple green, lilac

Element: Ba, Ca, K, Na

ii) For which group 2 compounds does the solubility in water (a) increase, and (b) decrease down the group?

The halogens

Key Term

Electronegative element
= one having a strong
affinity for an electron and
thus acting as an oxidising
agent.

These reactive, **electronegative elements** typically form anions having an oxidation state of –I so that oxidation is the usual reaction as in

$$Na(0) + \tfrac{1}{2}Cl_2(0) = Na^+(+I) + Cl^-(-I)$$

with the Na being oxidised and the oxidising chlorine being reduced from 0 to –I.

The tendency to form anions decreases down the group from fluorine to iodine with fluorine being the most electronegative element.

The melting temperatures of the elements increase down the group from gaseous fluorine to solid iodine owing to the increasing intermolecular forces holding the diatomic elements together in a liquid or solid. This increase is due to the increasing number of electrons in the molecules contributing to the induced dipole–induced dipole intermolecular force.

The halogens react with most metals to form halides with the reactivity decreasing down the group from fluorine to iodine. A similar feature is shown in displacement reactions in which a halogen higher in the group displaces one lower in the group from a salt as in

$$Cl_2 + 2NaBr = Br_2 + 2NaCl$$

This essentially reflects the decrease in oxidising power down the group with chlorine oxidising the bromide ion to bromine and being itself reduced to chloride

$$Cl_2(0) + 2Br(-I) = 2Cl^-(-I) + Br_2(0)$$

The reaction of halide ions with silver ions in dilute nitric acid is important in qualitative analysis in both organic and inorganic chemistry. The general reaction is

$$Ag^+(aq) + X^-(aq) = AgX(s)$$

The precipitate colours are chloride (white), bromide (pale cream) and iodide (pale yellow) and only the silver chloride dissolves in dilute ammonia. This gives a simple way of identifying the halogen present.

Grade boost

Be able to explain clearly why chlorine displaces bromine from bromides and bromine displaces iodine from iodides in terms of the decrease in oxidising power down the group.

quickfire

⑤ Write a balanced equation giving oxidation states for the action of chlorine on a solution of KBr.

Unit CH2 Summary Topics 4–6

Atoms are bonded together in molecules either by **covalent** bonds (electron pair sharing) or **ionic** bonds (electron transfer to form cations and anions). Ionic bonding gives a solid **giant lattice** with every cation surrounded by several, e.g. six, anions and vice versa, covalent bonding can either give **giant molecules**, e.g. diamond, where the atom forms four bonds to neighbours, or small molecules, e.g. H_2. Many bonds are **polar**, i.e. somewhere between covalent and ionic, if the bond atoms have different **electronegativities** (electron-attracting powers).

Covalent and ionic bonds are strong but there are weak intermolecular forces between molecules, usually **van der Waals** forces of two types; dipole–dipole if the molecules are polar, and induced dipole–induced dipole between all molecules. Also molecules containing hydrogen and very electronegative elements such as N, O and F form **hydrogen bonds**. These are stronger than usual van der Waals forces but still much weaker than covalent and ionic bonds.

The shapes of covalent molecules are governed by repulsions between the electron pairs around the central atom, **VSEPR**, that keep as far away from one another as possible giving linear (2 pairs), trigonal (3 pairs), tetrahedral (4 pairs), etc.; also lone pairs repel more than bond pairs and change the symmetrical shapes slightly.

Many elements are **metals** and here each atom gives up one or more electrons to form a lattice of positive ions held together by a sea of delocalised electrons; such a bond is usually strong.

Solubility of molecules in solvents depends on the type of intermolecular force with 'like dissolves like' being a useful rule. Polar, ionic and hydrogen-bonded molecules will tend to dissolve in the polar solvent water, whereas those having only van der Waals forces such as organic compounds will not.

The **Periodic Table** provides a useful guide to the trends in the properties of the elements relating to their electronic structures, showing how the key factors of ionisation energy, oxidation states, electronegativity, etc., change across periods and down groups and how these affect chemical behaviour. Here the detailed focus is on the chemistry of the s-block and the halogen elements.

In the **s-block** IEs are low, giving positive oxidation states, formation of cations and ionic salts, with the ease of cation formation showing that they are reducing agents giving their electron(s) to another species, such as chlorine, and themselves being oxidised to the cation.

The reverse is true of the **halogens**, where IEs are high, oxidation states negative, anions are formed in ionic salts in the process of oxidation where an electron is removed from, e.g. Na, and the halogen reduced to X^-.

7.1 Organic compounds and their reactions

Historically organic compounds were derived from living species but nowadays the term is applied to most carbon-containing compounds.

Hydrocarbon = a compound of carbon and hydrogen only.

Functional group = the atom/group of atoms that gives the compound its characteristic properties.

Saturated compound = one that contains no C to C multiple bonds.

Unsaturated compound = one that contains C to C multiple bonds.

Naming organic compounds

There are millions of possible organic compounds and each one has a specific name. To name a compound you have to know the homologous series to which it belongs – generally the functional group it contains.

The homologous series in this unit are:

alkanes – saturated hydrocarbons

alkenes – unsaturated hydrocarbons with a C to C double bond

halogenoalkanes – compounds in which one or more hydrogens in an alkane have been replaced by a halogen

primary (1°) alcohols – compounds containing –OH as the **functional group**

carboxylic acids – compounds containing the $-\overset{\displaystyle O}{\overset{\displaystyle \|}{C}}-OH$ (sometimes written as –COOH or $-CO_2H$) as the functional group.

You also need to know the 'code' that applies to the number of carbon atoms. In this:

meth = 1 carbon

eth = 2 carbons

prop = 3 carbons

but = 4 carbons

pent = 5 carbons

hex = 6 carbons

hept = 7 carbons

oct = 8 carbons

non = 9 carbons

dec = 10 carbons.

Rules

1. Find the longest continuous carbon chain. Using the code on p61, this is the basis of the name.
2. Number the C atoms in the chain, starting from the end that gives any side chains or substituted groups the smallest numbers possible.
3. If there is more than one side chain or substituted group the same, use the prefix di for 2, tri for 3 and tetra for 4.
4. Keep the alphabetical order of branch name.

It is easier to see how the rules work by applying them to particular examples.

2-methyl butane

Note: the $-CH_3$ group is called methyl.

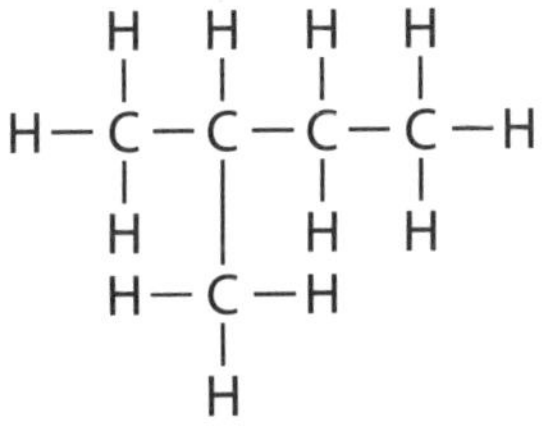

2,2-dimethyl propane

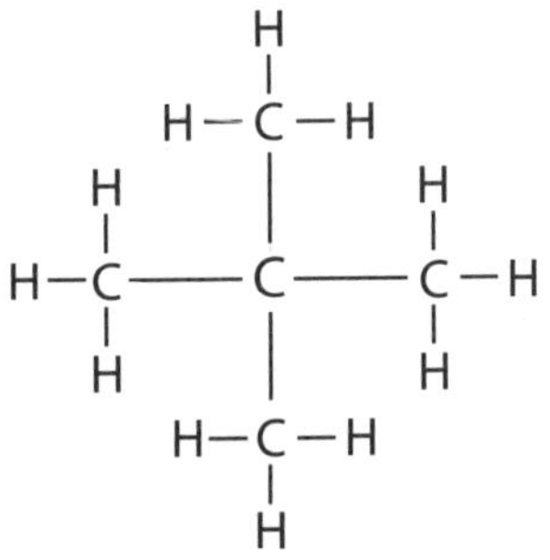

pent-1-ene,3-ol

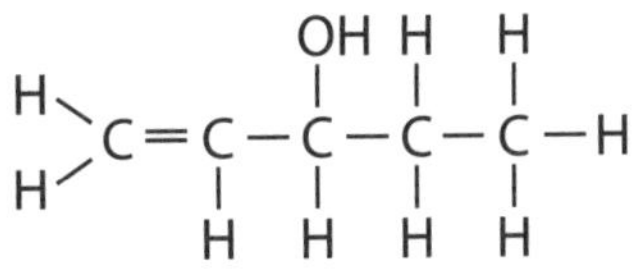

Note: $-$ OH is the functional group of alcohols and therefore the name includes $-$ ol.

C=C is the functional group for alkenes and therefore the name includes $-$ene.

Grade boost

A primary alcohol has the OH attached to a C attached to not more than 1 other C BUT, for this unit, it is the only type of alcohol you will be asked about.

Grade boost

The longest C chain can be bent/go round corners.

 quicKfire

① Find and number the longest C chain in the compound below.

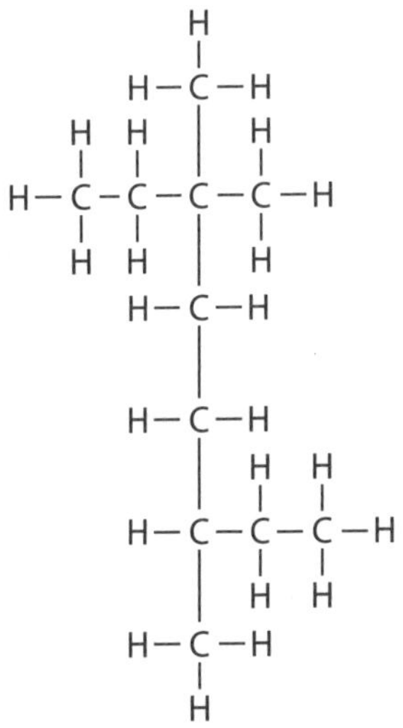

What is the 'code' that applies to the name of this compound?

Types of formulae

The formula of a particular compound can be shown in several different ways.

Molecular formula: shows the atoms, and how many of each type there are, in a molecule of compound.

Displayed formula: shows all the bonds and atoms in the molecule.

Shortened formula: shows the groups in sufficient detail that the structure is unambiguous.

Skeletal formula: shows the carbon/hydrogen backbone of the molecule as a series of bonds with any functional groups attached.

The different types can be seen using pent-1-ene,3-ol.

Molecular formula: $C_5H_{10}O$

Displayed formula:

```
H\              OH H  H
 C=C—C—C—C—H
H/  |  |  |  |
    H  H  H  H
```

Shortened formula: $CH_2CHCH(OH)CH_2CH_3$

Skeletal formula:

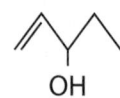

Homologous series

As described earlier, each set of compounds considered above belongs to a particular homologous series. An homologous series is a set of compounds that:

1. can all be represented by a general formula;
2. differ from their neighbour in the series by CH_2;
3. have the same functional group and so very similar chemical properties;
4. have physical properties that vary as the M_r of the compound varies.

The general formula of the alkane homologous series in C_nH_{2n+2} where n is an integer.

The effects of the functional group and the way in which physical properties, particularly melting and boiling temperatures, vary within an homologous series are considered later.

Empirical formulae

Methods used to analyse organic compounds often give results as masses or percentages of the elements present. These can be used to determine the empirical formula.

Example using masses

0.205 g of a hydrocarbon was burnt completely in oxygen and 0.660 g of carbon dioxide and 0.225 g of water were formed. The M_r of the compound was approximately 80.

(i) Calculate the empirical formula.

Mass of carbon in CO_2 = 12/44 × 0.660 = 0.180 g

Mass of hydrogen in H_2O = 2/18 × 0.225 = 0.025 g

$$
\begin{aligned}
\text{Ratio of number of moles} \quad & \text{C} : \text{H} \\
= \quad & \frac{0.180 : 0.025}{12.0 \quad 1.01} \\
= \quad & 0.0150 : 0.0248 \\
\text{Divide by smaller} = \quad & 1 : 1.65 \\
= \quad & 3 : 5
\end{aligned}
$$

Empirical formula is C_3H_5

(ii) What is the molecular formula of the hydrocarbon?

Since the molecular formula can be the same as, or a multiple of, the empirical formula, it is necessary to use its M_r.

M_r of empirical formula = 41

M_r of compound approximately 80 and therefore molecular formula is C_6H_{10}

Example using percentages

Analysis of a compound gave the percentage composition C = 60.0%, H = 13.3%, O = 26.7%. The M_r of the compound was 60. Determine its molecular formula.

$$
\begin{aligned}
\text{Ratio} \quad & \text{C} : \text{H} : \text{O} \\
= \quad & \frac{60.0 : 13.3 : 26.7}{12.0 \quad 1.0 \quad 16.0} \\
= \quad & 5.00 : 13.3 : 1.67 \\
= \quad & 3 : 8 : 1
\end{aligned}
$$

Empirical formula = C_3H_8O

Empirical M_r = 60, therefore molecular formula = C_3H_8O.

Key Term

Empirical formula = the formula of a compound with the atoms of the elements in their simplest ratio.

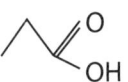

quickfire

⑤ Name the compound whose skeletal formula is shown below:

Grade boost

Always show **clearly** that you have calculated the M_r for your empirical formula.

quickfire

⑥ What is the general formula of the alkene homologous series?

⑦ What is the molecular formula of the alkane with 72 carbons?

⑧ A molecule has M_r of approximately 188. Its percentage composition is: C = 12.78; H = 2.15; Br = 85.07.
Find its empirical and hence its molecular formula.

Grade boost

Do not approximate when you 'divide by the smallest'. 1.67, for example, is $1\,^2/_3$. In this case you should multiply by 3 (to give 5).

Isomerism

Key Term

Structural isomers = compounds with the same molecular formula but a different structural formula, i.e. arrangement of the atoms.

Grade boost

You will be expected to recognise, name and draw isomers but those involving different functional groups are considered in a later unit.

quickfire

⑨ Draw displayed formulae for structural isomers of pentane. Name the isomers.

⑩ Draw skeletal formulae for two structural isomers of pentene that involve different positions of the double bond. Name the isomers.

Structural

Structural isomerism can arise in several ways:

1. **Chain isomerism** when the carbon chain is arranged differently.

    Example

    $CH_3CH_2CH_2CH_3$ butane

    and

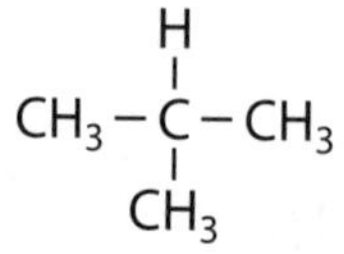

    $$CH_3 - \underset{\underset{CH_3}{|}}{\overset{\overset{H}{|}}{C}} - CH_3$$

    2-methylpropane

2. **Position isomerism** when the functional group is in a different position.

    Example

    $CH_2ClCH_2CH_3$ 1-chloropropane

    $CH_3CHClCH_3$ 2-chloropropane

3. **Functional group isomerism** when the functional group is different.

    Example

    $CH_3CH_2OCH_3$ an ether

    $CH_3CH_2CH_2OH$ propan-1-ol

E-Z isomerism

The single bond in alkanes allows free rotation but the double bond in alkenes means that rotation is restricted. This is due to the π bond – the reason for this is described later.

Since the double bond does not rotate, compounds such as 1,2-dibromoethene can exist in two different forms, i.e. two different isomers.

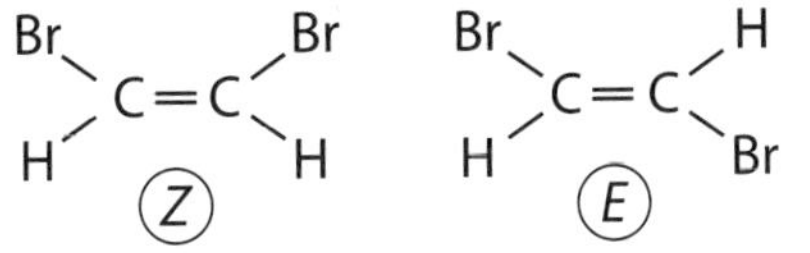

Key Term

E-Z isomerism = isomerism that occurs in alkenes (and substituted alkenes) due to restricted rotation about the double bond.

Naming E-Z isomers

To decide which isomer is which, you must look at the nature of the atom directly attached to each of the carbon atoms in the double bond. Look first at each carbon separately. The atom with the higher atomic number takes precedence.

Then look at how the higher priority groups are arranged – if they are both on the same side of the double bond, it is the Z isomer and if they are on opposite sides, it is the E isomer.

Example

For the isomers of 1,2-dibromoethene shown above:

on each side of the double bond Br has the higher A_r and therefore precedence;

in the LH structure the bromines are on the same side of the double bond, i.e. this is the Z form.

Properties of E-Z isomers

Since the functional groups are held differently, E-Z isomers can have different physical and chemical properties.

In, for example (Z) and (E) -butenedioic acids, the 2 acid groups can interact with each other in the Z form but are too far apart in the E form – this affects chemical reactions. They will also pack together differently and this affects physical properties – particularly melting and boiling temperatures.

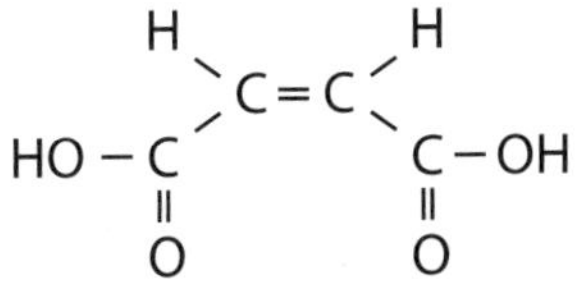

(Z)–butenedioic acid 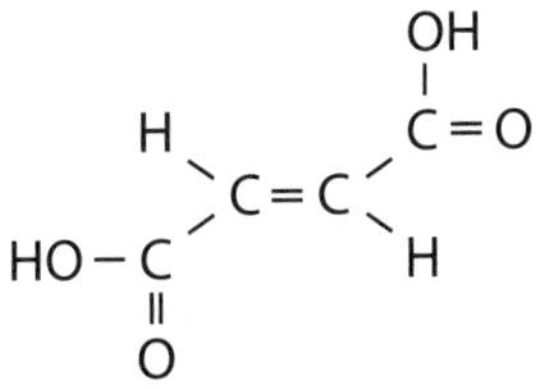   *(E)–butenedioic acid*

Grade boost

In older text books you may see reference to *cis-trans* isomerism. This has now been superseded by E-Z.

quickfire

⑪ Draw the structure to show (E)-1-chloro, 2-bromoethene.

⑫ Does the structure below show the E or Z isomer?

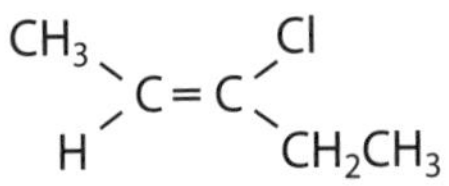

Grade boost

If you need a mnemonic to remember which is which, Z comes from the German zusammen = together and E from entgegen = opposite.

7.2 Hydrocarbons

The effect of chain length on melting and boiling temperatures

Key Term

Van der Waals forces = dipole–dipole or induced dipole–induced dipole interactions between atoms and molecules.

Grade boost

Look back at the section on Van der Waals forces and make sure you can explain the formation of induced dipole–induced dipole forces.

quickfire

⑬ The boiling temperature of pentane is 36°C. Suggest a value for the boiling temperature of hexane.

⑭ Draw structural formulae for pentane and 2,2-dimethylpropane and use these to explain why the boiling temperature of pentane is 36°C whilst that of 2,2-dimethylpropane is 10°C.

When a simple covalent substance melts or boils, heat energy is supplied. This is to overcome Van der Waals forces. Which substance has a higher melting or boiling temperature can be predicted by looking at the strength of these forces.

For hydrocarbons only induced dipole–induced dipoles are present and these are weak. Since the intermolecular forces happen at the surface, it is necessary to look at the surface areas of the molecules. A small surface area means that lower Van der Waals forces are possible and this means that the melting and boiling temperature would be low.

Small hydrocarbons are therefore gases at room temperature, larger ones are liquids and even larger ones are solids.

You know that the boiling temperature of compounds increases with increasing chain length. If different structural isomers are considered, they have different surface areas. This can be used to explain boiling temperatures.

The more branches an isomer has, the more like a sphere it is and the lower is the boiling temperature.

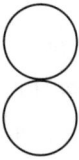

many branches – little surface area contact

straight chain – more surface area contact

Petroleum

Fractional distillation

Since **petroleum** is a mixture of many hydrocarbons, before it can be used, it must be separated into fractions. This is done using a fractionating column.

Each of the fractions has a different boiling temperature range so that the lowest boiling temperature compounds reach the top of the column, whilst the highest boiling temperature compounds remain at the bottom.

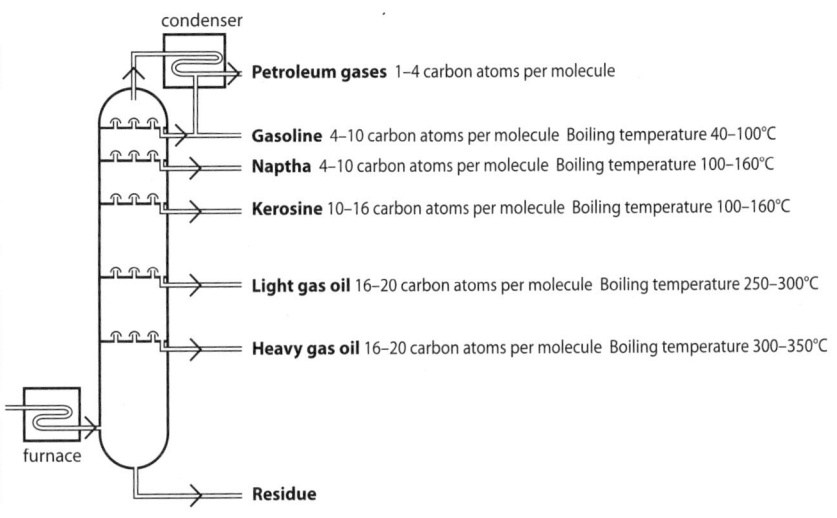

condenser

Petroleum gases 1–4 carbon atoms per molecule

Gasoline 4–10 carbon atoms per molecule Boiling temperature 40–100°C

Naptha 4–10 carbon atoms per molecule Boiling temperature 100–160°C

Kerosine 10–16 carbon atoms per molecule Boiling temperature 100–160°C

Light gas oil 16–20 carbon atoms per molecule Boiling temperature 250–300°C

Heavy gas oil 16–20 carbon atoms per molecule Boiling temperature 300–350°C

furnace

Residue

The different fractions can be collected at various points up the column.

Cracking

The fractions produced by **fractional distillation** do not correspond to the uses for which hydrocarbons are needed. Much more petrol is needed than is available from this fraction. The larger, higher boiling temperature fractions are less useful and so these are cracked to give smaller, more useful molecules.

Cracking can be achieved using heat (thermal cracking) but it takes less energy to use a catalyst (catalytic cracking). When hydrocarbons are cracked, the bonds can break in many ways, so a variety of products will form but the conditions of the cracking plant can be adjusted to maximise the yield of the most desirable products. Cracking can give straight, branched chain or cyclic alkanes and alkenes as well as hydrogen.

Key Terms

Fractional distillation = separating into fractions according to boiling temperatures.

Petroleum = a complex mixture of hydrocarbons.

Grade boost

A fraction still contains a mixture of a number of hydrocarbons but they now all have similar boiling temperatures.

quickfire

⑮ Would petroleum extracted from beneath the North Sea give the same fractions as petroleum from Kuwait?

⑯ Why must petroleum be separated into fractions before it is used?

Grade boost

You do not need to be able to quote exact numbers of carbons in each fraction or its boiling temperature but you should recognise that lower M_r fractions are used in petrol, whilst heavier fractions are used as diesel and lubrication oil, and the heaviest are used to produce bitumen for roads.

quickfire

⑰ Write the equation to show decane being cracked to give hexane and one other product.

⑱ Why are alkanes generally so unreactive?

Grade boost

There are only three mechanisms in this unit. At least one of them is asked on nearly every Unit 2 exam and so you should make sure you really understand them!

⑲ Why are radicals so reactive?

⑳ How is C_2H_6 formed as one product in the reaction between methane and chlorine?

㉑ Write the overall equation for the formation of CCl_4 from CH_4.

Grade boost

The stages (initiation, etc.) are the mechanism. A question may ask for this or the overall equation such as:
$C_2H_6 + Cl_2 \rightarrow C_2H_5Cl + HCl$

Photochlorination of methane

Although alkanes are generally unreactive, they will react with halogens in the presence of uv light (often as sunlight). The mechanism of this reaction is in three stages.

Stage 1 – initiation

The energy needed to homolytically break the bond in chlorine is provided by uv light.

$$Cl_2 \rightarrow 2Cl^{\bullet}$$

$Cl^{\bullet}$ is then a radical.

Stage 2 – propagation

Radicals are very reactive and will take part in a series of propagation reactions.

$$Cl^{\bullet} + CH_4 \rightarrow CH_3^{\bullet} + HCl$$
$$CH_3^{\bullet} + Cl_2 \rightarrow CH_3Cl + Cl^{\bullet}$$

In the propagation stages each reaction starts with a radical and then produces one so that the chain reaction continues.

Stage 3 – termination

The chain reaction continues until two radicals meet in a termination stage.

$$CH_3^{\bullet} + Cl^{\bullet} \rightarrow CH_3Cl$$

Further substitution

Since radicals are so reactive other propagation stages can occur, e.g.

$$CH_3Cl + Cl^{\bullet} \rightarrow CH_2Cl^{\bullet} + HCl$$
$$CH_2Cl^{\bullet} + Cl_2 \rightarrow CH_2Cl_2 + Cl^{\bullet}$$

This means that polysubstitution can occur and a mixture of products is formed.

Overall, this type of mechanism is described as being radical substitution.

Ethene and the alkenes

Structure and bonding

Alkenes contain a carbon to carbon double bond. In ethene the double bond consists of a sigma (σ) bond and a pi (π) bond. The π bond is formed from the sideways overlap of a p orbital electron on each carbon.

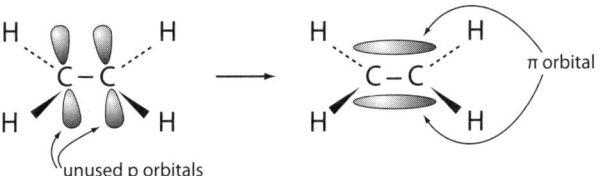

Reactions

Mechanism: electrophilic addition

The pair of electrons in the π orbital means alkenes are susceptible to attack by an **electrophile**. The mechanism involves heterolytic bond fission and leads overall to electrophilic **addition**.

If the attacking species is not polar, a dipole is induced by the negative charge of the π bond.

X_2 could be H_2 or Br_2.

Key Terms

Electrophile = an electron deficient species that can accept a lone pair of electrons.

Addition reaction = a reaction in which reagents combine to give one product.

Grade boost

When drawing a mechanism you must show exactly where the 'curly arrows' start and finish. The arrow generally starts from a lone pair or a π bond.

quickfire

22 Why is the mechanism of addition to alkenes said to involve heterolytic bond fission?

Grade boost

Be careful when to use + and when to use $\delta+$. + is produced when a neutral species loses an electron, $\delta+$ forms as part of a dipole.

Grade boost

In describing a test, always include the appearance before and after the test.

quickfire

㉓ Why, in the manufacture of 'butter substitute' spreads, is nickel generally used as a catalyst (rather than platinum or palladium)?

㉔ Complete the equation:
$(CH_3)_2C=CH_2 + HBr \rightarrow$

㉕ Draw the structure for a 3° carbocation. (You can choose the number of carbon atoms.)

Uses of the reaction

Test for alkenes

Uses bromine as a test for alkenes since the brown colour of bromine changes to colourless (aqueous bromine is generally preferred since it is safer).

Unsaturated compounds to saturated compounds

Uses hydrogen to convert double bonds to single bonds in the presence of a catalyst (Pt, Pd or Ni). Used to convert unsaturated oils and fats to saturated fats – harder and used as 'butter substitutes'.

Which product?

Unsymmetrical alkenes could add HBr to give two different products.

$$H_2C=CH-CH_3 + HBr \longrightarrow CH_3-CHBr-CH_3$$

propene   +HBr   2-bromopropane

$$H_2C=CH-CH_3 + HBr \longrightarrow CH_2Br-CH_2-CH_3$$

propene   +HBr   1-bromopropane

The major product is 2-bromopropane due to the greater stability of the 2° **carbocation** (compared with the 1° carbocation).

2° carbocation   1° carbocation

Preparation of alkenes

Alkenes can be prepared from halogenoalkanes by an **elimination reaction**. Hydrogen halide is removed (an acid) and therefore an alkali is used. Halogenoalkane vapour is passed over heated soda lime or they are dissolved in ethanol and heated with sodium hydroxide.

$$CH_3CH_2Br + NaOH \rightarrow CH_2=CH_2 + NaBr + H_2O$$

Polymerisation of alkenes and substituted alkenes

Polymerisation is the joining of a very large number of **monomer** molecules to make a large polymer molecule.

Alkenes, and substituted alkenes, undergo **addition polymerisation**. In this type of polymerisation the double bond is used to join the monomers and nothing is eliminated.

Ethene is polymerised to make poly(ethene) (commonly called polythene).

The names of **polymers** still include the name of the monomer but the polymers do not contain double bonds.

Poly(ethene) is unreactive and flexible so it is used to make plastic bags, etc.

When poly(ethene) was first made, the polymer chains had side branches coming from the main chain and these prevented the polymer chains packing together. This meant that the density of the polymer was low and there were few points of contact for Van der Waals forces so that the melting temperature was also low.

Catalysts can be used to make poly(ethene) with straight chains. This means that the chains can pack together so that the polymer has a higher density and higher melting temperature. These properties mean that it is used where more rigidity is needed and/or the temperature is comparatively high.

The properties of polymers can also be altered by using substituted alkenes as the monomer. This means that these polymers have a huge variety of uses.

Note: You do not need to be able to quote specific uses for specific polymers but you should be aware of the principles involved in the polymerisation and how the different physical properties make different polymers suitable for different functions.

Grade boost

When drawing a section of a polymer chain you must show — at both ends to show that the chain continues.
If you only draw one **repeat unit**, do not forget to put 'n' outside the bracket.

quickfire

26 What is the name of the polymer formed from 1-chloro, 2-cyanoethene?

27 Draw two repeat units for the polymer you have named in QF26.

28 What is the empirical formula of poly(ethene)?

Other widely-used polymers

These include:

Poly(propene)

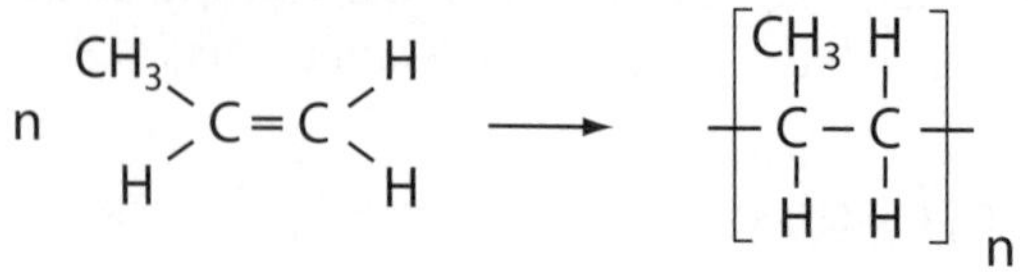

monomer
propene

polymer
poly(propene)

Poly(propene) is rigid and used in food containers and kitchen equipment.

Poly(chloroethene)

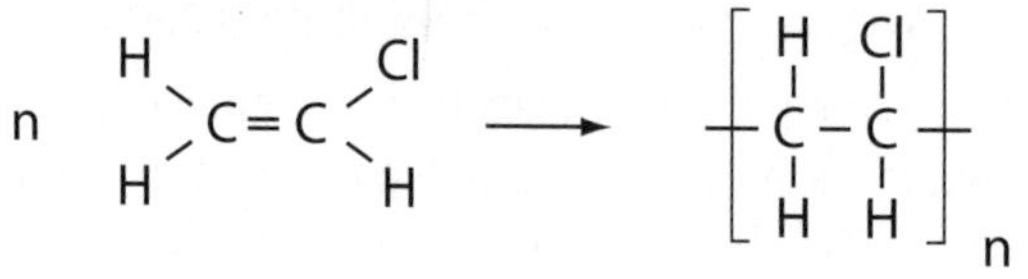

monomer
chloroethene

polymer
poly(chloroethene)

Polychloro(ethene) used to be called PVC and its properties can be modified to make it a flexible covering for electrical cable insulating covering as well as pipes, etc.

Poly(phenylethene)

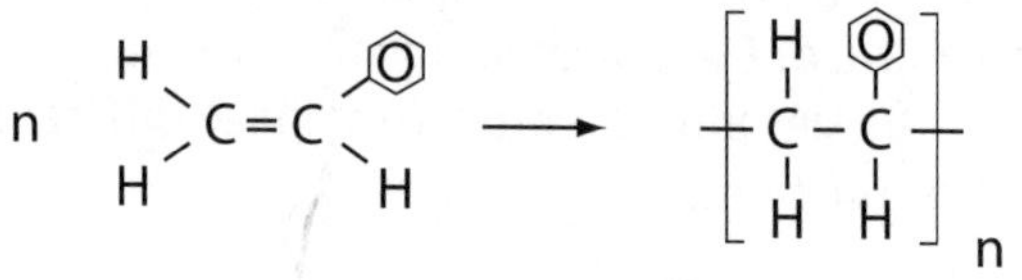

monomer
phenylethene

polymer
poly(phenylethene)

Poly(phenylethene) used to be called polystyrene and, since it is hard, it is used in many household items needing strength and rigidity. It can be made into an insulator by creating holes in the structure (expanded polystyrene).

Grade boost

An easy way of drawing the formula of a polymer when you are told the monomer is to draw the monomer with all substituted groups around the double bond i.e.

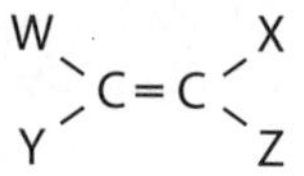

quickfire

㉙ A section of polymer is shown below. Draw the monomer that was used to form this polymer.

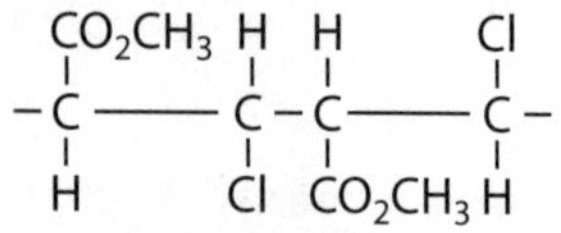

㉚ Suggest two changes that could be made to the structure of a polymer to lower its melting temperature.

7.3 Halogenoalkanes

Formation

They are formed when halogens react with alkanes. See Photochlorination of methane for details and mechanism.

Structure

Since halogens are more electronegative than carbon, **halogenoalkanes** are polar. The halogen is $\delta-$ and the carbon attached to the halogen is $\delta+$.

Example is bromomethane

$$H - \overset{\overset{\displaystyle H}{|}}{\underset{\underset{\displaystyle H}{|}}{C}} \overset{\delta+}{-} \overset{\delta-}{Br}$$

This $\delta+$ means that the carbon is electron deficient and hence susceptible to **nucleophilic** attack.

Hydrolysis

Water is a **nucleophile** since there are two lone pairs on the oxygen. Halogenoalkanes can be **hydrolysed** by water, e.g.

$$C_2H_5I + H_2O \rightarrow C_2H_5OH + HI$$
iodoethane   ethanol

However, this reaction is rather slow so that a more powerful nucleophile, OH^-, is generally used. This is provided by NaOH(aq).

The mechanism for the reaction is:

$$C_3H_7\overset{\overset{\displaystyle H}{|}}{\underset{\underset{\displaystyle H}{|}}{C}} \overset{\delta+}{-} \overset{\delta-}{Cl} \longrightarrow C_3H_7\overset{\overset{\displaystyle H}{|}}{\underset{\underset{\displaystyle H}{|}}{C}} - OH + Cl^-$$

:OH$^-$

1- chlorobutane   chlorobutane-1-ol

The nucleophile attacks the $\delta+$ carbon, donates a lone pair and forms a bond to the carbon. The carbon to chlorine bond breaks to give the chloride ion.

Overall this reaction is nucleophilic **substitution** with 1-chlorobutane producing butan-1-ol.

Key Terms

Halogenoalkanes = a homologous series in which one or more hydrogen atoms in an alkane has been replaced by a halogen.

Hydrolysis = a reaction with water to produce a new product.

Nucleophile = a species with a lone pair of electrons that can be donated to an electron-deficient centre.

Substitution reaction = a reaction in which one atom/ group is replaced by another.

Grade boost

The hydrolysis of 1-chlorobutane is one of the mechanisms required in the specification. Examiners like to ask questions that show you understand mechanisms!

quickfire

㉛ Draw the displayed formula of 1,2-dichlorobutane.

㉜ Draw the skeletal formula of 3-iodopent-2-ene.

㉝ Is reacting chlorine with ethane a satisfactory method of preparing pure 1-chloroethane? Explain your answer.

Grade boost

The use of aqueous silver nitrate in the test for the presence of a halogen in an organic compound is the same as you used in Unit 1 to test for halide ions in inorganic compounds.

quickfire

㉞ Write down, in order, the steps needed to test for the presence of bromine in an organic compound. Include the result expected.

㉟ Write the ionic equation for the reaction that occurs to produce the yellow precipitate formed if iodine is present in an organic compound.

Test for the presence of a halogen in an organic compound

Since the hydrolysis of an organic compound containing a halogen produces a halide ion, the reaction can be used to show the functional group – X (where X = Cl, Br or I).

In practice, the organic compound is heated with aqueous sodium hydroxide.

$$RX + NaOH(aq) \rightarrow ROH + Na^+(aq) + X^-(aq)$$

R is the alkyl group (or other organic part of the molecule).

The presence of X^- (aq) can be shown by adding $AgNO_3$(aq) but any NaOH(aq) remaining after the hydrolysis would interfere with this test and must be removed. This is done by adding HNO_3(aq).

The presence of the halogen, in the original organic compound, can then be seen in the usual test for halides.

halogen	addition of Ag^+(aq)	addition of NH_3(aq) to precipitate formed with Ag^+(aq)
chlorine	white precipitate	dissolves in dilute NH_3(aq)
bromine	cream precipitate	dissolves in concentrated NH_3(aq)
iodine	yellow precipitate	does not dissolve in NH_3(aq)

Solubility of halogenoalkanes

Although halogenoalkanes are polar, they do not contain the O–H or N–H needed to form hydrogen bonds with water. They are therefore insoluble in water but they are able to mix with a variety of non-polar/polar organic substances. This means they can be used as solvents in a variety of processes and that they are very effective degreasing agents. This includes their use in dry cleaning. Chlorocompounds are most commonly used as they are the cheapest.

Other uses of halogenoalkanes

Although small halogenoalkanes are gases at room temperature the presence of permanent dipole–permanent dipole attractions means that boiling temperatures are close to room temperature. They are therefore liquids that can easily be evaporated or gases that can easily be liquefied at room temperature.

This, and their comparative unreactivity, led to a variety of uses. These included:

1. Fire extinguishers used tetrachloromethane (CCl_4) as a liquid whose vapour could suppress a fire.

2. Refrigerants used chlorofluorocarbons (**CFCs**). Heat is needed to change a liquid to a gas and this heat is removed from the fridge to cool its contents.

3. Propellants used CFCs. When the pressure on a liquefied CFC is released, it changes to a gas. This, as it escapes from its container, brings the active ingredient (hair spray/ fly spray, etc.) with it.

4. Anaesthetic – for example tri-chloromethane ($CHCl_3$) was an early form of an anaesthetic. This was called chloroform.

Environmental effects of CFCs

CFCs undergo radical chain reactions to destroy the Earth's **ozone layer**. Since this absorbs UV radiation that can cause skin cancer, it is undesirable.

Initiation stage

Initiation occurs due to bond fission of the C–Cl bond in CFC to produce radicals. This is caused by uv radiation in the upper atmosphere. It is the C–Cl bond, rather than the C–H or C–F bond, that is broken because this is the weakest of these bonds.

Using trichlorfluoromethane as an example of a CFC.

$$CFCl_3 \rightarrow Cl^{\bullet} + CFCl_2^{\bullet}$$

Propagation stage

The process is complex and there are many possible propagation reactions. These include:

$$Cl^{\bullet} + O_3 \rightarrow ClO^{\bullet} + O_2$$
$$ClO^{\bullet} + O_3 \rightarrow Cl^{\bullet} + 2O_2$$

This is a chain reaction and so the formation of a small number of chlorine radicals can cause the decomposition of many ozone molecules.

Due to problems of ozone depletion, much work has been done to find replacements for CFCs. Suggestions include the use of hydrofluorocarbons (**HFCs**) since these do not contain C–Cl bonds and cannot give chlorine radicals.

Organohalogen compounds as pesticides and polymers

Many pesticides used in the past contained **organohalogen compounds**. These included DDT which built up in animal food chains and caused lowering in the numbers of animals at the top of the chain.

The polymer poly(chloroethene) (PVC) is still widely used but when it is burnt it produces hydrogen chloride, as well as the usual carbon monoxide and carbon dioxide produced in the combustion of organic material. This hydrogen chloride gives an acidic atmosphere.

Key Terms

HFCs = halogenoalkanes containing fluorine as the only halogen.

Organohalogen compounds = a wide variety of generally complex organic molecules containing halogens.

Ozone layer = a layer around the earth containing O_3 molecules.

quickfire

36 How can CFCs be liquefied so that they can be used as refrigerants or propellants?

Grade boost

The C–Cl bond, rather than the C–F bond, breaks because fluorine is smaller than chlorine and therefore forms stronger covalent bonds.

quickfire

37 How do you recognise a propagation step in a reaction mechanism?

38 Write the overall equation for the propagation steps shown in the destruction of the ozone layer.

Grade boost

Do not try to remember the propagation steps shown in the main text. Many alternatives are possible but you should recognise that overall you are changing O_3 to O_2 whilst regenerating the chlorine radical.

Grade boost

Hydrogen bonds are a particularly strong form of intermolecular force and so have more effect on melting and boiling temperatures than other dipole-dipole interactions.

quickfire

(39) Why is the boiling temperature of pentan-1-ol higher than that of propan-1-ol?

(40) The boiling temperature of ethane is −89°C, that of chloroethane is 12°C and that of ethanol is 78°C. Explain the difference in these boiling temperatures.

Grade boost

When drawing hydrogen bonds to explain boiling temperatures or solubilities, be careful that you use the correct species and that you mark the hydrogen bonds between the δ^+ and the lone pair.

7.4 Alcohols

Alcohols contain the functional group –OH. Compounds with more than one OH do exist but in this unit you only need to know about those with just one.

Structure

In the OH group O is more electronegative than H and it has lone pairs. This means that alcohols are polar and can form **hydrogen bonds**.

Boiling temperatures

Since alcohols can form hydrogen bonds between their molecules, more energy has to be supplied to overcome these and therefore alcohols have higher boiling temperatures than corresponding hydrocarbons.

As with other homologous series, as the M_r increases the boiling temperature also increases, but for alcohols even methanol is a liquid at room temperature.

Using R to be an alkyl group the relatively high boiling temperature can be explained using a diagram.

Shows hydrogen bond between alcohol molecules.

The boiling temperature of the commonest alcohol, ethanol, is about 80°C. Since a high boiling temperature implies low **volatility**, alcohols are less volatile than would be expected from their M_r.

Solubility in water

The OH in the alcohol can form hydrogen bonds with the OH in water so that small alcohols are soluble in water.

The industrial preparation of ethanol

Ethene reacts with steam to produce ethanol.

$$CH_2=CH_2(g) + H_2O(g) \rightleftharpoons CH_3CH_2OH(g) \quad \Delta H = -45 \text{ kJ mol}^{-1}$$

The conditions used are a temperature of 300°C, a pressure of 60–70 atmospheres and a catalyst of phosphoric acid (coated onto an inert solid). The conditions used can be explained using **Le Chatelier's principle**.

Temperature

The forward reaction is exothermic and so a high yield would be favoured by a low temperature. This, however, would give a slow rate of reaction, so 300°C is a compromise temperature.

Pressure

Two moles of gaseous reactants give one mole of gaseous product and so a high yield is favoured by a high pressure. This also increases the rate of reaction but too high a pressure is expensive to maintain.

Catalyst

This increases the rate of the reaction without affecting the yield. Using these conditions gives about 5% conversion of ethene to ethanol and therefore the unreacted ethene is recycled back to the reaction chamber.

Dehydration of primary alcohols

Many alcohols can be dehydrated to form alkenes. The equation for butan-1-ol is shown below.

$$CH_3CH_2CH_2CH_2OH \rightarrow CH_3CH_2CH=CH_2 + H_2O$$
$$\text{but-1-ene}$$

A variety of dehydrating agents can be used but the most commonly used ones are heated aluminium oxide or concentrated sulfuric acid.

Since water is removed (the H from one carbon atom and the OH from the next) and a double bond is formed, this is an example of an **elimination reaction**.

Key Terms

Elimination reaction = a reaction that involves the removal of 2 atoms/ groups from adjacent atoms in a molecule to produce a multiple bond.

Le Chatelier's principle Look back to Unit 1 and make sure you understand this principle.

Grade boost

The reaction between ethene and steam is electrophilic addition. It has the same mechanism as the other reactions of alkenes.

 quickfire

41 From where might the ethene, needed to make ethanol, be obtained on a large scale?

42 How could ethanol be removed from the equilibrium mixture produced when ethene reacts with steam?

Grade boost

The dehydration of alcohols is the reverse of the reaction of alkenes with steam. You will need to be able to quote a suitable dehydrating agent and recognise that others are possible.

Heat under reflux = to heat liquids in a flask fitted with a vertical condenser so that the liquid evaporates, then condenses and falls back into the flask.

Grade boost

If you include the oxidation number of, e.g. potassium dichromate (VI), it must be correct. If you are unsure, it might be safer to omit the number.

quickfire

㊸ Write equations to show:
(a) dehydration
(b) oxidation of propan-1-ol.

Grade boost

You may see $Cr_2O_7^{2-}/H^+$ in a reaction scheme. This shows the oxidising agent acidified dichromate(VI).

When you use [O] in an oxidation equation, do not forget to balance the equation.

quickfire

㊹ What would you expect to see, as a positive result, if a breathalyser based on acidified dichromate(VI) were used?

Oxidation of primary alcohols

Many alcohols can also be oxidised. The usual oxidising agent is acidified potassium dichromate(VI). The alcohol, potassium dichromate(VI) and sulfuric acid are **heated together under reflux**. The fact that oxidation has happened is shown since the dichromate(VI) changes from orange to green.

Using [O] to represent the oxidising agent and R to be an alkyl group, the equation shows oxidation to produce a carboxylic acid.

$$R-\overset{\overset{\displaystyle H}{|}}{\underset{\underset{\displaystyle H}{|}}{C}}-OH \ + \ 2[O] \ \longrightarrow \ R-C\overset{\displaystyle O}{\underset{\displaystyle OH}{}} \ + \ H_2O$$

The alcohol in alcoholic drinks is ethanol and if these drinks are exposed to air some oxidation to ethanoic acid occurs. This acid is present in vinegar and therefore the drink would taste 'sour' and unpleasant.

Drinks containing ethanol

Alcoholic drinks contain ethanol and, although for some people this is an addictive drug, drinking alcohol is acceptable in many cultures. Many social functions are based on alcohol consumption, since people feel relaxed when they drink small quantities of ethanol. There is evidence indeed that small quantities of alcohol are beneficial to health but consuming large quantities, over a long period, causes significant health problems – particularly affecting liver function.

Alcohol slows reaction times and therefore it is dangerous to drink and drive. A legal limit for alcohol in the bloodstream is in force. The amount of alcohol a person had drunk used to be tested by blowing through a breathalyser based on acidified potassium dichromate(VI) but nowadays methods based on the IR spectrum of ethanol are used.

Topic 7 Summary and examples of some important terms in organic reactions

In this unit you have seen and used all the following terms. Make sure you understand them and can recognise them when they occur.

Addition reaction

A reaction in which reagents combine to give one product

Example

$CH_2 = CH_2 + Br_2 \rightarrow CH_2BrCH_2Br$

Substitution reaction

A reaction in which one atom/group is replaced by another

Example

$C_2H_5Br + OH^- \rightarrow C_2H_5OH + Br^-$

Elimination reaction

A reaction that involves the removal of 2 atoms/groups from adjacent atoms in a molecule to produce a multiple bond

Example

$C_2H_5OH \rightarrow CH_2 = CH_2 + H_2O$

Homolytic bond fission

Breaking a bond so that each atom involved in the bond receives one of the bond's electrons

Example

$Cl_2 \rightarrow 2Cl^•$

Heterolytic bond fission

Breaking a bond so that one of the atoms involved in the bond receives both of the bond's electrons

Example

$C_2H_5Br + OH^- \rightarrow C_2H_5OH + Br^-$

Electrophile

An electron-deficient species that can accept a lone pair of electrons

Example

$CH_2 = CH_2 + Br_2 \rightarrow CH_2BrCH_2Br$
Br_2 with δ^+ is the electrophile

Nucleophile

A species with a lone pair of electrons that can be donated to an electron-deficient centre

Example

$C_2H_5Br + OH^- \rightarrow C_2H_5OH + Br^-$
OH^- is the nucleophile

Radical

A species with an unpaired electron

Example

$Cl_2 \rightarrow 2Cl^•$
$Cl^•$ is a radical

Oxidation

A change involving loss of electrons BUT in organic reactions often reaction with [O]

Example

$C_2H_5OH + 2[O] \rightarrow CH_3COOH + H_2O$

Hydrolysis

A reaction with water to produce a new product

Example

$C_2H_5Br + OH^- \rightarrow C_2H_5OH + Br^-$

Grade boost

Rearrangements of parts of the molecule can occur. Do not try to explain the identity of all the peaks in a mass spectrum in an exam. You will be asked about those that help give the structure.

quickpire

① Why are spectroscopic techniques generally preferred to volumetric and gravimetric methods nowadays?

② What information is given by the y axis values in a mass spectrum? Is this useful in finding the structure of the compound?

Grade boost

Subtraction of m/z values can show what has been broken off – for example, in the pentane mass spectrum $72 - 57 = 15$ shows the loss of CH_3.

8 Analytical techniques

Traditionally to analyse the nature and quantity of an unknown substance, volumetric and gravimetric analysis techniques were used. In volumetric analysis, titrations are used to measure volumes of solutions, whilst in gravimetric analysis, masses of solids are found. Nowadays spectroscopic techniques are more often the preferred method.

In this unit you need to be able to analyse mass and IR spectra to help identify the structure of an organic molecule. In later units you will study other spectroscopic methods.

Mass spectrometry

In a mass spectrometer an electron is knocked off molecules of an organic compound to produce a positive ion. This is the **molecular ion** and is often shown as M^+. The mass spectrometer also causes the molecules to split into smaller parts – **fragments**. These fragments can give information about the structure of the molecule.

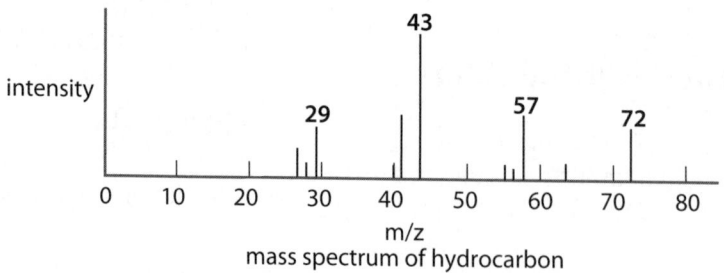

mass spectrum of hydrocarbon

The x axis in a mass spectrum shows mass/charge (m/z) but you can assume that the charge on the ion is 1. To interpret a mass spectrum, look at the peak with the largest m/z. This is the molecular ion and it gives the M_r.

For this hydrocarbon $M_r = 72$ and, as it contains only C and H this suggests a molecular formula of C_5H_{12} i.e. an isomer of pentane.

Then look at specified fragments. $29 = C_2H_5^+$, $43 = CH_3CH_2CH_2^+$ and $57 = CH_3CH_2CH_2CH_2^+$.

This suggests the compound is $CH_3CH_2CH_2CH_2CH_3$.

If chlorine or bromine are present in the compound, two peaks for M^+ and some of the fragments will be seen since both halogens can exist as two isotopes.

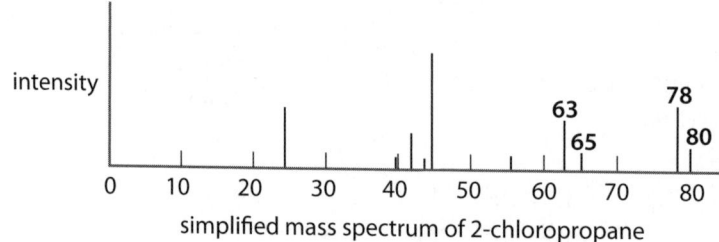

simplified mass spectrum of 2-chloropropane

From the spectrum, M^+ peak is at 78 for $C_3H_7{}^{35}Cl$ but at 80 for $C_3H_7{}^{37}Cl$. The peaks at 63 and 65 are due to the loss of 15 i.e. $CH_3{}^+$ from the molecule.

The other fragment peaks are caused by rearrangements and are difficult to interpret from the structure of 2-chloropropane.

Infrared spectroscopy

Radiation in the IR part of the spectrum is absorbed to cause increased vibrations and bending in organic molecules.

The wavenumber at which the **absorption** occurs is **characteristic** of the bond and therefore useful in identifying the functional group present.

You will be given the wavenumbers you need to answer exam questions. Examples of these are shown in the table.

Bond	Wavenumber /cm⁻¹
C–C	650 to 800
C–O	1000 to 1300
C=O	1650 to 1750
C–H	2800 to 3100
O–H	2500 to 3550

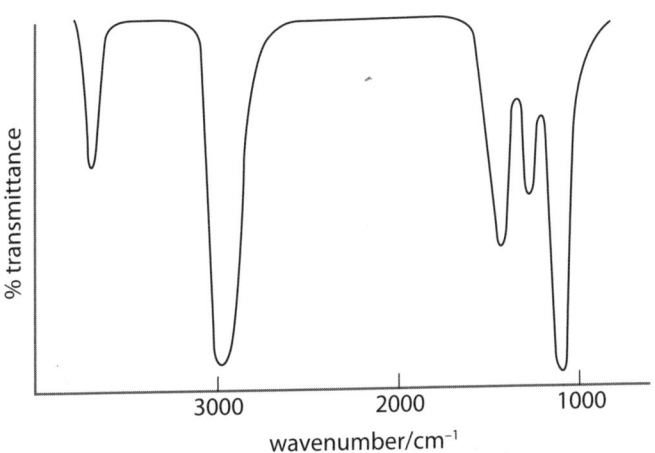

infrared spectrum of ethanol

In the same way that not all peaks were useful in mass spectra, not all absorptions are useful in interpreting IR spectra. Just look at the information you have and look for absorptions corresponding to the functional groups/bonds present.

The spectrum of ethanol shown above is consistent with the structure of ethanol since it shows an absorption at approximately 1050 cm⁻¹ from C–O and one at approximately 3000 cm⁻¹ from O–H.

It would have been difficult, without some information from another source, to categorically state O–H was present. The peak could have been caused by the presence of C–H.

Grade boost

It is the troughs, i.e. peaks downwards, that are used in IR spectra.

quickfire

③ Why are the peaks at 78 and 80 not the same height in the mass spectrum of 2-chloropropane?

④ At what m/z would you expect to see the largest value for propan-1-ol?

What group has been lost from propan-1-ol to give a peak with m/z at 43?

⑤ Why is an absorption in the range 2800 to 3100 not really useful in determining the groups present?

⑥ In which of the compounds

$CH_3C \overset{O}{=} OCH_3$

and CH_3OCH_3 would you expect to see an absorption at 1700 to 1720 cm⁻¹?

Grade boost

The presence, or absence, of a peak at about 1700 cm⁻¹ is really useful. It shows C=O and, if present, it is always very sharp and clear.

Exam Practice and Technique

Exam practice and skills

WJEC AS Chemistry aims to encourage students to:

- develop their interest and enthusiasm for the subject, including developing an interest in further study and careers in the subject
- appreciate how society makes decisions about scientific issues and how the sciences contribute to the success of the economy and society
- develop and demonstrate a deeper appreciation of the skills, knowledge and understanding of How Science Works
- develop essential knowledge and understanding of different areas of the subject and how they relate to each other.

Examination questions are written to reflect the assessment objectives as laid out in the specification. Candidates must meet the following assessment objectives in the context of the content detailed in the specification.

Assessment objective AO1: Knowledge and understanding of science and How Science Works

Candidates should be able to:

- recognise, recall and show understanding of scientific knowledge
- select, organise and communicate relevant information in a variety of forms.

37% of the questions set on the exam paper are recall of knowledge.

Assessment objective AO2: Application of knowledge and understanding of science and How Science Works

Candidates should be able to:

- analyse and evaluate scientific knowledge and processes
- apply scientific knowledge and processes to unfamiliar situations including those relating to issues
- assess the validity, reliability and credibility of scientific information.

37% of the questions set on the AS exams are AO2.

Assessment objective AO3: How Science Works

Candidates should be able to:

- demonstrate and describe ethical, safe and skilful practical techniques and processes, selecting appropriate qualitative and quantitative methods
- make, record and communicate reliable and valid observations and measurements with appropriate precision and accuracy
- analyse, interpret, explain and evaluate the methodology, results and impact of their own and others' experimental and investigative activities in a variety of ways.

26% of the questions set on the AS exams are on How Science Works.

CH1 and CH2: Written paper (1 hour 30 minutes each)

The papers have two sections.

Section A consists of about six short answer objective questions adding to 10 marks. Section B consists of structured questions, normally five, and totals 70 marks giving a total of 80 marks for each paper.

NB. CH3 is a practical coursework exam.

How exam questions are set

The examiner sets questions that must fall within the Specification and provide a fair balance of the various sections. The provisional paper and the mark scheme are checked by a reviser and an evaluation committee to ensure that the paper is error-free and of the correct standard.

How exam questions are marked

Answered papers are marked by a panel of examiners after the provisional mark scheme has been thoroughly tested by trial marking both before and during a standardising meeting. Each examiner then marks exactly according to the finally agreed scheme. If an unexpected answer turns up, the marker contacts the principal examiner for a ruling. The marking team use their professional judgement in most borderline cases. The general policy is that an answer that has both correct and relevant chemistry will be awarded the mark.

The questions are worded very carefully so that they are clear, concise and unambiguous. Despite this, candidates tend to penalise themselves unnecessarily when they misread questions, either because they read them too quickly or too superficially. It is essential that candidates appreciate the precise meaning of each word in the question if they are to be successful in producing concise, relevant and unambiguous responses. The mark value at the end of each part of each question provides a useful guide as to the amount of information required in the answer.

Valuable resources that your teacher will help you with are the Specification itself, copies of past papers to show what the questions are actually like and annual reports from the examiner covering the strengths and weaknesses of the candidates.

Exam tips

Read the question carefully. Examiners try to make the wording of the questions as clear as possible but in the examination situation, it is all too easy to misinterpret a question. Read every word in every sentence carefully and use a highlighter pen if it helps you to focus on key words.

Understand the information

Only a certain percentage of the questions at AS are based on recall of knowledge. You may encounter unfamiliar material. It is important that you do not panic but think carefully and take your time and apply the principles that you have learned to answer these types of question. You many also encounter a graph or table. Again, read the information carefully several times before you attempt the question.

Look at the mark allocation

Each question or part of a question is allocated a number of marks. You must make sure that if the question is worth three marks, then you must give three points to gain those marks.

Understand the instructions

Know the meaning of action words. Make sure that you are familiar with the terms below and that you understand what the examiner expects you to do.

Describe

This term may be used in a variety of questions where you need to give a step-by-step account of what is taking place. In a graph question, for example, you may be required to recognise a simple trend or pattern, then you should also use the data supplied to support your answer. At this level it is insufficient to state that the graph goes up and then flattens. You are expected to describe what goes up and by how much.

Explain

A question may ask you to describe and also explain. You will not be given a mark for merely describing what happens – a chemical explanation is also needed.

Suggest

This action word often occurs at the end of a question when you are expected to put forward a sensible idea based on your chemical knowledge. There may not be a definite answer to this question.

Name

This means that you must give no more than a one-word answer. You do not have to repeat the question or put your answer into a sentence. This is wasting time.

State

A brief, concise answer with no explanation.

Compare

If you are asked to make a comparison, do so. For example, if you are asked to compare aldehydes and ketones, don't write out two separate descriptions.

Tips about structured questions

Structured questions can be short, requiring a one-word response, or can include the opportunity for extended writing. The number of lined spaces on the exam question paper together with a mark allocation are indications of the length of answer expected and the number of points to be made. Structured questions are in several parts, usually about a common content. There is an increase in the degree of difficulty as you work your way through the question. The first part may be simple recall, perhaps defining a term, the most difficult part coming at the end of the question.

Improving performance

- Know the specification.
- Work on past papers.
- Do enough calculations.
- Give yourself enough time to absorb and understand the subject; last-minute swotting is no good. Your subconscious will go on working for you if you give it time.
- Read the questions slowly and carefully, do not overlook a section in your haste.
- Don't leave sections blank, your guess may be worth a mark.
- Give enough time to check back over your answers, especially calculations.

- Are your numerical answers sensible?
- The material in one section of a question may give a clue as to the answer in another part; look up and down though part questions.

Where students go wrong

- You should assume that the examiner is an idiot, spell everything out, leave no uncertainty.
- Don't say 'it' use the name.
- Leave no doubt in the examiner's mind, e.g. in 'describe a test to distinguish between chloride and bromide ions in dilute nitric acid', the answer has three parts: 1 – the test to use, 2 – the result for chloride, and 3 – the result for bromide. Students often say 'the chloride precipitate dissolves in dilute ammonia' without mentioning that the bromide doesn't.
- Words in bold in the questions require exact answers, e.g. name means name only, a formula will be marked wrong, three significant figures means that giving more or fewer figures will be wrong, e.g. 58.5 – correct, 58 – wrong, 58.52 – wrong. Also displayed, shortened and skeletal structural formulae must be exactly correct when in bold.
- Questions may say 'what is observed'. The answer must be what you would see, such as 'bubbles of gas formed' not 'carbon dioxide is produced', or 'a white precipitate is formed' not 'silver chloride is produced'.

Finally

Sports men and women and musicians who want to be good have to work and practise hard; it is the same for you. Sitting round in a group in the library chatting a bit and idly turning over your notes is not work, it is passing the time.

To do well you need to get a good set of notes, concentrate on them until you know them, keep testing back to check that you do know them and apply your knowledge to past and practice questions. None of us is keen on work but through work comes an enjoyment of mastery of the subject and a confidence that you will do well.

Good luck!

Questions and answers

This part of the guide looks at student answers to examination-style questions through the eyes of an examiner. There is a selection of questions on topics in the AS specification with two sample answers – one of a high grade standard and one of a lower grade standard in each case. The examiner commentary is designed to show you how marks are gained and lost so that you understand what is required in your answers.

CH1: Controlling and Using Chemical Changes

CH2: Properties, Structure and Bonding

Q & A 1

Radiation

In April 1986 there was an accident at the nuclear plant in Chernobyl. The table below shows the names, symbols, half-lives and types of emission of some of the radioactive isotopes that were found in the clouds and rain over the Welsh hills soon after the accident.

Element	Symbol of isotope	Half-life of isotope	Type of emission
Caesium	$^{137}_{55}Cs$	28 years	β
Iodine	$^{131}_{53}I$	8 days	β
Ruthenium	$^{106}_{44}Ru$	374 days	β

(i) State what happens in the nucleus of an atom when a ß-particle is emitted. *[1]*

(ii) Give the mass number and symbol of the product of the radioactive decay of ruthenium. *[1]*

(iii) I Explain why radioactive isotopes may be a health hazard. *[2]*

   II State why there is more concern about radioactive contamination due to the isotope $^{137}_{55}Cs$ than $^{131}_{53}I$. *[1]*

(iv) I Explain what is meant by the term *half-life* of a radioactive isotope. *[1]*

   II Calculate how long it would take for 2.0g of $^{106}_{44}Ru$ to be reduced to 0.25g of $^{106}_{44}Ru$. *[1]*

Tom's answer

(i) A neutron decays. ✗ ①

(ii) Mass number 103, symbol Rh ✗ ②

(iii) I Radiation causes cell mutation ✓ this leads to cancer. ✓

   II The radioactivity of the isotope exists for a longer time. ✓

(iv) I The time taken for half the radioisotopes in a sample to decay. ✓

   II 2 g to 0.25 g decreases by factor of 4, so time taken is 374 x 4 = 1496 days. ✗ ③

Examiner commentary

① Tom has not been specific enough; he needs to state that a neutron decays to form a proton and an electron.

② Tom has given the correct symbol but has used the Periodic Table to find the mass number. The Periodic Table does not give the mass number of a particular isotope.

③ Dividing the initial mass by the final mass does not give the number of half-lives.

Tom achieves 4 out of 7 marks: Grade C.

Seren's answer

(i) A nucleus loses an electron. ✓ ①

(ii) Mass number 106, symbol Rh. ✓

(iii) I Radiation is ionising ✓ this produces radiation burns. ✓ ①

   II It has a longer half-life. ✓ ②

(iv) I Time taken for half of the atoms in a sample of a radioisotope to decay. ✓

   II 1122 days. ✓

Examiner commentary

① These are acceptable alternatives to the answers given by Tom.

② This is acceptable since the question only says 'state', it does not ask for an explanation.

Seren achieves 7 out of 7 marks: Grade A.

Q&A

2

Ionisation energies

The first and second ionisation energies of potassium and sodium are shown in the table below.

	1st ionisation energy/kJ mol^{-1}	2nd ionisation energy/kJ mol^{-1}
Potassium	419	3051
Sodium	496	5059

(a) Explain the term *molar first ionisation energy*. [2]

(b) Explain why:
  (i) potassium has a lower first ionisation energy than sodium [2]
  (ii) there is a large difference between the first and second ionisation energies of potassium. [2]

Tom's answer

(a) It is the energy required to remove an electron from an atom to create an ion. ✗ ①

(b) (i) Because its outermost electron (4s¹) is further away from the positively charged nucleus. ✓ Also in potassium the 4s¹ orbital is shielded by the 3p orbital. ✗ ②

  (ii) Because for the second ionisation energy an electron is removed from potassium's 3p shell which is full; however, for the first an electron is in a shell by itself so requires less energy to remove. ✗ ③ Also it's further away from the attraction of the nucleus. ✓

Examiner commentary

① Tom gains no marks as he has not made any reference to the mole and he has not mentioned the gaseous state.

② Tom does not gain the second mark because, although he has given a correct statement, he has not made a valid comparison.

③ Although Tom's statement is correct, it is too vague. He needs to state why an electron in a full shell requires more energy to remove it than an electron in a shell by itself.

Tom achieves 2 out of 6 marks: Grade D.

Seren's answer

(a) It is the energy required to remove one mole of electrons from one mole of its gaseous atoms. ✓ ✓

(b) (i) The outer electron in potassium is a further distance from the nucleus ✓ and there is increased shielding in potassium. ✓ ①

  (ii) The first electron is removed from the 4s shell and the second electron is removed from the 3p shell which is closer to the nucleus. ✓ ②

Examiner commentary

① Seren has provided all the points required to gain both marks. She could also have stated that sodium has a greater effective nuclear charge.

② Seren only gains one mark because both statements cover the same marking point. To get the second mark she needs to state that the shielding effect on the outer electron is less or the effective nuclear charge is greater after the first electron has been removed.

Seren achieves 5 out of 6 marks: Grade A.

Atomic hydrogen spectrum

(a) Describe the visible emission spectrum of atomic hydrogen and explain how its features relate to electronic levels within the hydrogen atom. [4]

(b) Explain how the ionisation energy of hydrogen can be derived from the Lyman series in the atomic hydrogen spectrum. [3]

Tom's answer

(a) The spectrum is a pattern of separate lines. ✓ ① Because the lines are separate, it shows that the energy levels in the atom are quantised. ✓ ②

(b) The energy lines converge to a limit as they go from $n = 1$ to $n = \infty$ and this represents the ionisation energy. ✓ ③

Examiner commentary

① Tom gains one mark for the description. To gain the second mark he needs to describe the pattern that the lines form.

② Tom gets one mark for correctly using 'quantised'; however, he has not explained how the lines form.

③ Tom has successfully stated the relationship between ionisation energy and the Lyman series but he has not explained how the ionisation energy can be derived.

Tom achieves 3 out of 7 marks: Grade D.

Seren's answer

(a) The visible emission spectrum of atomic hydrogen is a series of lines ✓ which become closer as the energy increases. ✓ The lines are caused by excited electrons dropping back to a lower energy level ✗ ① therefore electrons exist in discrete energy levels and energy levels are quantised. ✓

(b) For the Lyman series, $n = 1$, the convergence limit represents the ionisation of the hydrogen atom. ✓ The convergent frequency (difference from $n = 1$ to $n = \infty$) can be measured ✓ and the ionisation energy can be calculated from $\Delta E = hf$. ✓

Examiner commentary

① Seren has an excellent grasp of this topic and has given a full answer. However, in the visible emission spectrum (Balmer series) the lines are the result of electronic transitions from higher energy levels to energy level $n = 2$ and Seren must state this.

Seren achieves 6 out of 7 marks.

Q&A 4

Mass spectrometer

The mass spectrometer is an important analytical instrument.
State the functions of the parts of the instrument labelled A to F below. *[6]*

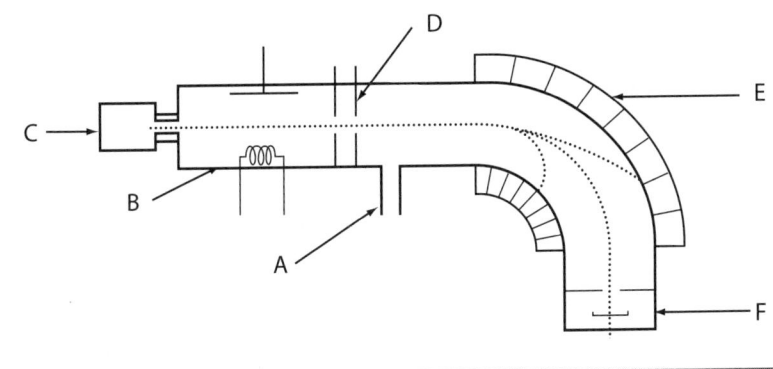

Tom's answer

A Vacuum pump ✗ ①
B Heated element producing electrons ✓
C Heats the sample ✗ ②
D Speeds up the ions ✓
E Deflects the ions ✗ ③
F Detects the ions ✓

Examiner commentary

① Tom has named the part and not given the function. If he had stated 'to remove air' he would have gained the mark.

✓ ② Although Tom's statement is correct, it is not specific enough. The sample must be gaseous when it enters the mass spectrometer.

③ Although the magnetic field deflects the ions, its function is to separate them by deflecting the ions according to their mass/charge ratio.

Tom achieves 3 out of 6 marks: Grade C.

Seren's answer

A To stop air particles from colliding with sample particles ✓
B Ionisation chamber to produce ions ✗ ①
C Vaporisation of the sample ✓
D Accelerates the ions ✓
E Separates the ions ✓
F Amplifies the signal ✓ ②

Examiner commentary

① Seren needs to state that positive ions are formed in the chamber.

② This is acceptable because after the ions have been detected, the signal is amplified and recorded.

Seren achieves 5 out of 6 marks: Grade A.

Q&A 5

Moles

Baking powder contains mainly sodium hydrogencarbonate.

4.75 g of baking powder are contained in 250cm³ of solution. A 25.0cm³ portion of the aqueous solution required 29.9cm³ of 0.170 mol dm⁻³ aqueous hydrochloric acid solution for neutralisation.

Sodium hydrogencarbonate reacts with aqueous hydrochloric acid according to the equation:

$$NaHCO_3 + HCl \rightarrow NaCl + H_2O + CO_2$$

  (i)   Calculate the number of moles of hydrochloric acid used.   [1]

  (ii) Deduce the number of moles of sodium hydrogencarbonate in the 25.0 cm³ portion of baking soda.   [1]

 (iii) Calculate the number of moles of sodium hydrogencarbonate in the original 250 cm³ solution.   [1]

  (iv) Calculate the mass, in grams, of sodium hydrogencarbonate in the original baking powder solution.   [1]

  (v) Calculate the percentage by mass of sodium hydrogencarbonate in the baking powder.   [1]

  (vi) Calculate the volume of carbon dioxide that would be produced at 25°C.   [1]
          (1 mole of carbon dioxide occupies 24 dm³ at 25°C.)

Tom's answer

(i) Mol HCl = 0.17 x 29.9 = 5.08 ✗ ①

(ii) Mol NaHCO₃ = 5.08 ✓ ②

(iii) Mol NaHCO₃ in original solution = 5.08 x 10 = 50.8 ✓

(iv) Mass in solution = 50.8 x 83.01 = 4217 g ✗ ③

(v) % NaHCO₃ = $\frac{4.75}{4217}$ x 100 = 0.11% ✗

(vi) Volume CO₂ = 5.08 x 24 = 122 dm³ ✓ ②

Examiner commentary

✓

① Tom has forgotten to divide the volume by a thousand to change cm³ into dm³.

② Tom gets the mark due to consequential marking.

③ The M_r of NaHCO₃ is incorrect.

Tom achieves 3 out of 6 marks: Grade C.

Seren's answer

(i) Mol HCl = $\frac{29.9 \times 0.17}{1000}$ = 5.08 x 10⁻³ ✓

(ii) Mol NaHCO₃ = 5.08 x 10⁻³ ✓

(iii) Mol NaHCO₃ in original solution = 5.08 x 10⁻³ x 10 = 5.08 x 10⁻² ✓

(iv) Mass in solution = 5.08 x 10⁻³ x 84.01 = 4.27g ✓

(v) % NaHCO₃ = $\frac{4.27}{4.75}$ x 100 = 89.9% ✓

(vi) Volume CO₂ = 5.08 x 10⁻³ x 24 = 0.122 ✗ ①

Examiner commentary

① Seren needs to state the correct unit for the volume.

Seren achieves 5 out of 6 marks: Grade A.

Q & A 6

Equilibria

Ammonia is used in the industrial production of nitric acid. The production of nitrogen monoxide, NO, from ammonia involves the equilibrium:

$$4NH_3(g) + 5O_2(g) \leftrightarrows 4NO(g) + 6H_2O(g) \quad \Delta H = -909 \text{ kJ mol}^{-1}$$

(a) State the meaning of the term *dynamic equilibrium*.   [1]

(b) State and explain the effect on the proportion of nitrogen monoxide in the equilibrium mixture by:
    (i)   increasing the pressure   [2]
    (ii) increasing the temperature   [2]

(c) The catalyst used in the reaction above is an alloy of platinum and rhodium. Name this type of catalyst.   [1]

Tom's answer

(a) Dynamic equilibrium is when the forward reaction equals the backward reaction. ✗

(b) i) Increasing the pressure will shift the position of equilibrium to where there is least moles, i.e. to the left-hand side. ✓ ✗ ①

  ii) The equilibrium yield decreases ✓ as the equilibrium shifts to the left ✗ ②

(c) Metallic. ✗

Examiner commentary

① Although Tom's statement is correct, he has not answered the question fully – he has not stated the effect on the proportion of nitrogen monoxide. So he only gains one mark

② Tom needs to explain why the equilibrium shifts to the left to gain the mark.

Tom achieves 2 out of 6 marks: Grade D.

Seren's answer

(a) When the rate of the forward reaction equals the rate of the backward reaction. ✓

(b) i) The yield would decrease ✓ because there are fewer (gas) moles on the left-hand side. ✓

  ii) The forward reaction is exothermic ✓ ① so equilibrium shifts to the left to counteract the change and the yield of nitrogen monoxide decreases. ✓

(c) Heterogeneous. ✓

Examiner commentary

① Seren could have stated that the backward reaction is endothermic. In reversible reactions it is very important to state forward or backward reaction not just reaction.

Seren achieves 6 out of 6 marks: Grade A.

Q & A 7

Titration

Elinor determined the concentration of a solution of hydrochloric acid by titrating it against a standard solution of sodium carbonate. She rinsed the burette with acid, filled it to above the zero mark using a funnel, opened the tap and checked the burette jet. She removed the funnel and brought the acid level to exactly $0.00 cm^3$.

She placed the sodium carbonate solution in a conical flask with an indicator and added the acid while swirling the flask. When the indicator gave signs of change, she added the acid drop by drop to the end-point.

The readings of her titrations were 20.80, 20.20, 20.05 and $20.10 cm^3$ respectively.

(i) State why the burette was rinsed with the acid before filling. [1]

(ii) State why the jet of the burette was looked at. [1]

(iii) State why the funnel was removed. [1]

(iv) State and explain whether there was any need for the acid level to be set exactly on zero. [1]

(v) State why the flask was swirled during the titration. [1]

(vi) State why the acid was added drop by drop at the end. [1]

(vii) Identify any anomalous result and calculate a mean value for her titration. [1]

Tom's answer

(i) In case it was dirty. ✓

(ii) To ensure that the acid flows freely. ✓

(iii) For safety reasons. ✗

(iv) Yes, to ensure that the acid level was the same at the beginning of each titration. ✗

(v) To mix all the reactants. ✓

(vi) So that she would not add too much acid. ✓

(vii) Mean value = $\dfrac{20.80 + 20.20 + 20.05 + 20.10}{4}$ = $20.29 cm^3$ ✗ ①

Examiner commentary

① Tom has included all four results in his calculation for the mean value. The first result is significantly higher than the others; therefore it should not be included in the calculation.

Tom achieves 4 out of 7 marks: Grade C.

Seren's answer

(i) To ensure that it was clean. ✓

(ii) To check for air bubbles. ✓

(iii) To prevent any further drops of acid falling in. ✓

(iv) No, the result is the difference between the final and initial values. ✓

(v) To ensure that all the sodium carbonate reacts. ✓

(vi) Not to overshoot the end-point. ✓

(vii) Mean value $20.12 cm^3$ ✓ ①

Examiner commentary

① Although Seren has not specifically stated that $20.80 cm^3$ is an anomalous result, her answer for the mean titre shows that she has only included three titres in her calculation and so she gets the mark.

The question has been worded this way to indicate to the students that they should not use all the results in their calculation.

Seren achieves 7 out of 7 marks: Grade A.

Q & A

8

Enthalpy

Describe a laboratory experiment of your choice, for determining the enthalpy change of a reaction. Your answer should include details of the apparatus to be used, the measurements to be taken and the way in which you would use your results to determine the enthalpy change.

[6]

Tom's answer

To determine the enthalpy change for the neutralisation reaction between magnesium oxide and dilute hydrochloric acid.

Pour hydrochloric acid into a coffee cup, take the temperature then add a weighed amount of magnesium oxide. Take the temperature of the solution every 30 seconds for the next 6 minutes. ✓✓ ✗✗ ①

To find the enthalpy of the reaction use the expression $\Delta H = \dfrac{-mc\Delta T}{n}$

where m is the mass of the solution, c is a given constant, ΔT the maximum temperature change and n the number of moles of magnesium oxide. ✓ ✗ ②

Examiner commentary

① Although the method is correct, Tom's answer is not detailed enough. He needs to: State what volume of hydrochloric acid he's adding and what apparatus he uses to add it.

Ensure that the temperature of the acid is constant before he adds the magnesium oxide and state the accuracy of the thermometer.

Stir the reactants vigorously and ensure that the coffee cup has a lid.

② To obtain ΔT he needs to draw a graph of temperature against time and extrapolate to the beginning of the experiment. He also needs to make it clear that if he is using the number of moles of magnesium oxide then the acid is in excess.

Tom achieves 3 out of 6 marks: Grade C.

Seren's answer

I will determine the enthalpy change for the displacement reaction between zinc and copper(II) sulfate solution using the following method.

Pipette 50.0cm³ of copper(II) sulfate solution of known concentration into a polystyrene cup.

Put a thermometer through the hole in the lid and record the temperature to the nearest 0.1°C every half minute until the reading is constant. ✓✓

Weigh about 6g of zinc and add the powder to the cup. (The zinc is in excess so there is no need to be accurate.)

Stir the solution well and record the temperature every half minute until the temperature fall is constant. ✓✓

Draw a graph of temperature against time and extrapolate the curve to the beginning of the experiment to find the maximum temperature change.

Use $\Delta H = -mc\Delta T$ to find the enthalpy change for the experiment.

Scale to 1 mole by dividing by the number of moles of copper(II) sulfate. ✓✓ ①

Examiner commentary

① Seren has an excellent grasp of this topic and has given a full description of the experiment. She could have stated what the *m* and *c* stand for in the expression for ΔH. However, in a question of this type there will always be more marking points than marks available, therefore she can gain full marks without giving all the salient points.

Seren achieves 6 out of 6 marks: Grade A.

Q & A 9

Reaction rates (theory)

Discuss and explain the following by considering the movement and the energy of the particles involved.

Calcium carbonate reacts faster with:

(a) concentrated hydrochloric acid than it does with dilute hydrochloric acid [3]

(b) dilute hydrochloric acid at 60°C than at 20°C. [4]

Tom's answer

(a) When the hydrochloric acid is concentrated, there are more molecules present in a certain volume. ✓ This means there is more chance of collisions per unit time and so the reaction goes faster. ✓ ✗ ①

(b) The different temperatures mean that the particles in the acid will be moving at different speeds, the higher the temperature the higher the speed. ✓ If the particles are moving at greater speeds this means more collisions per second. A greater rate of collisions means a faster rate of reaction. ✗ ②

Seren's answer

(a) In concentrated hydrochloric acid there are more particles in a given volume ✓ therefore there will be more molecular collisions taking place, so there will be a greater frequency of collisions ✓ ① that actually cause a reaction. ✓ This makes the reaction go faster.

(b) In order to react, collisions must occur between particles with at least a minimum amount of energy – the activation energy. ✓ When the temperature of the acid increases, the kinetic energy of its particles also increases. ✓ This results in a larger proportion of the particles having enough energy to react ✓ so the frequency of successful collisions increase and the rate of reaction increases. ✓ ②

Examiner commentary

① Tom's answer gives the impression that all collisions cause a reaction – only molecules with sufficient energy react when they collide. This is a common mistake, be careful to avoid it.

② Tom's idea that there are more collisions because the reaction goes faster is not detailed enough to gain any more marks. He should have used the idea of activation energy – the minimum energy needed by a particle in order to react on collision – in his answer.

Tom achieves 3 out of 7 marks: Grade D.

Examiner commentary

① Frequency of collision is equivalent to collisions per unit time.

② Seren has given the full explanation expected at this level.

Seren achieves 7 out of 7 marks: Grade A.

Q&A

10

Reaction rates (practical)

Hydrogen peroxide, H_2O_2, decomposes to give water and oxygen.
An experiment was conducted to investigate the decomposition of 40cm³ of 0.40 mol dm⁻³ H_2O_2 solution at a constant temperature of 25°C in the presence of 0.5g of MnO_2 catalyst. The results are shown below:

Time/s	0	10	20	40	80	120	160	200	240
Volume O_2/cm³	0	12	24	40	75	89	94	96	96

(a) Outline a suitable method, including essential apparatus, for carrying out an experiment to obtain these results. [4]

(b) Use the results to plot a graph of the volume of oxygen against time, and from the graph calculate the initial rate of reaction. [5]

Tom's answer

(a) Using a measuring cylinder pour the H_2O_2 solution into a conical flask ✓ add the MnO_2 and measure the volume of oxygen formed until the reaction stops using a stopwatch. ✗ ①

(b)

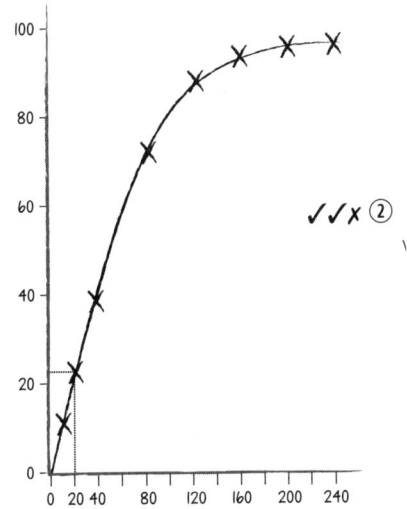

✓✓✗ ②

Initial rate = $\dfrac{volume}{time}$ = $\dfrac{40}{40}$ = 1.0cm³ s⁻¹ ✗ ✓ ③

Examiner commentary

① Tom would score 1 mark for pouring the H_2O_2 solution into a suitable reaction vessel, but the rest of the method is not detailed enough to score any marks.

② Tom has not labelled the axes and so loses 1 mark.

③ Tom has not used his graph correctly. The line begins to curve after 20 s so he needs to measure the gradient of the line between 0 and 20 s.

Tom achieves 4 out of 9 marks: Grade D.

Seren's answer

(a) A measuring cylinder would be used to pour 40cm³ of the H_2O_2 solution into a conical flask. ✓ 0.5 g of the MnO_2 would be weighed accurately. The reaction is done at room temperature, which should stay constant, during the reaction. ✗ ①

When ready, the MnO_2 would be added quickly to the H_2O_2 solution and a bung and delivery tube attached to a gas syringe inserted into the flask ✓ and a stopwatch started. At appropriate intervals the volume of the oxygen in the syringe would be recorded. ✓

(b)

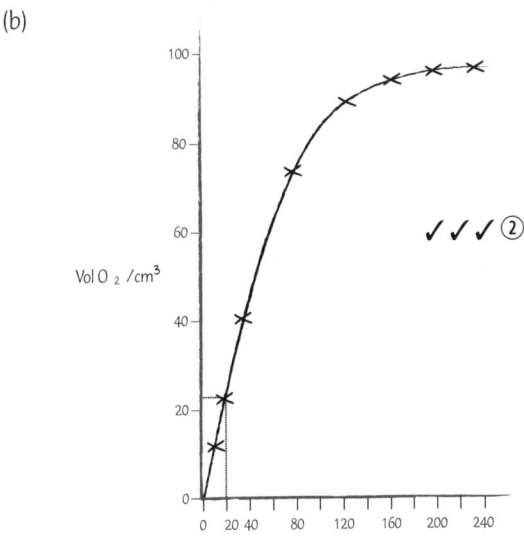

✓✓✓ ②

Initial rate = $\dfrac{volume}{time}$ = $\dfrac{24}{20}$ = 1.2 ✓ ✗ ③

Examiner commentary

① Seren's choice of apparatus is good and this gains her 3 marks. However, her method does not control temperature. To gain another mark she should have immersed her flask in a water bath measured at 25°C.

② Seren has drawn her graph accurately, labelled the axes suitably and drawn a smooth curve so gains all 3 marks.

③ Seren has not given any units, so loses 1 mark.

Seren achieves 7 out of 9 marks: Grade A.

Q&A 11

Bonding

(a) Use 'dot and cross' diagrams to show:
   (i) Covalent bonding in Li_2,
   (ii) The formation of an ionic bond with LiCl.
   (Only outer electrons need to be shown.) [3]
(b) Explain what is meant by a **co-ordinate bond** and give an example. [3]
(c) List the forces of electrical attraction and repulsion that exist in a diatomic covalent molecule. [3]

Tom's answer

(a) (i) Li $\overset{\circ}{\times}$ Li   ✓ ①

(ii)

$$Li^{\times} \; ^{\circ}Cl^{\circ}_{\circ} \longrightarrow Li \quad Cl$$ ✓ ✗ ②③

(b) A bond in which both electrons come from one of the atoms. ④

$$H \; N \; H^{+} \quad ✗✓ ⑤$$
H

(c) Electron–electron repulsion, ✓ ⑥
Electrons attracted to both nuclei ✓ ⑦
⑧ ✗

Examiner commentary

① Correct drawing of Li_2
② Electron transfer OK
③ The formation of the ion with charges must be shown.
④ Description OK but limited
⑤ Not clear that both electrons come from one atom.
⑥⑦ OK
⑧ Nucleus – nucleus repulsion omitted.
Tom gained 5 marks out of 9: Grade C.

Seren's answer

(a) (i) Li $\overset{\circ}{\times}$ Li   ✓ ①

(ii)

$$Li^{\times} \; ^{\circ}Cl^{\circ}_{\circ} \longrightarrow Li^{+} \quad Cl^{-}$$ ✓✓

(b) A covalent-type bond where one of the atoms provided both electrons in the electron bond pair. ✓ Such a bond will be polar to some extent. ✓ ②

$$H \; N \; H^{+}$$
H   ✓

(c) Electron-electron repulsion ✓
Electrons attracted to both nuclei ✓
Nucleus- nucleus repulsion ✓ ③

Examiner commentary

① Correct example used
② Full answer and accurate diagram.
③ Correct – repulsion between the positive nuclei is often forgotten.
Seren obtained full marks: Grade A.

Q&A 12

Electronegativity

(a) Explain what is meant by the term electronegativity and show how electronegativity values are useful when considering bond polarity. *[4]*

(b) Using the electronegativity values below:

  (i) Arrange the covalent bonds given in order of INCREASING polarity C–Cl; C–H; K–H; O–H. *[2]*

  (ii) Write the δ+ and δ- charge signs of each atom in the bonds. *[2]*

Electronegativity values: K 0.8, H 2.1, C 2.5, Cl 3.0, O 3.5

Tom's answer

(a)This is a measure of the attraction of an atom in a covalent bond to the electron pair in the bond.✓ ✓ ①

Values are useful since the more electronegative elements will give more polar bonds ✗ ②

(b) (i) C–H< C–Cl< O–H< H–K ✓ ✗ ③

(ii) δ + δ-   δ+ δ-   δ+ δ-   δ- δ+

    C--Cl   C---H   K--H   O- -H   ✓ ✗ ④

Examiner commentary

①② Definition OK, but it is the difference in electronegativity that is important not the actual value, e.g. F-F is nonpolar.

③ H–K should be before O–H

④ The C should be δ- in C–H

Tom obtained 4 marks out of 8: Grade D.

Seren's answer

(a) A measure of the ability of an atom in a covalent bond to attract the electron pair bond.✓ ✓ The values are useful since the difference between the two values for the atoms in the bond is proportional to the bond polarity. ✓ ✓ ①

(b) (i) C-H<C-Cl<H-K<O-H   ✓ ✓ ②

  (ii) δ + δ-   δ- δ+   δ+ δ-   δ- δ+

     C---Cl   C---H   K--H   O- -H   ✓ ✓ ③

Examiner commentary

① All correct here, 4 marks awarded.

② Correct, so 2 marks given.

③ Correct so another 2 marks given.

Seren gained the full 8 marks: Grade A.

Van der Waals bonding

(a) State which **one** of the following bonds is generally the weakest:

      covalent   hydrogen   ionic   van der Waals   *[1]*

(b) (i) Describe the nature of van der Waals forces and explain the difference between the two types of van der Waals force. *[4]*

    (ii) For each of these types above name a molecule where the force is important. *[2]*

(c) State **one** effect that van der Waals forces have on the physical properties of compounds. *[1]*

Tom's answer

(a) Van der Waals ✓ ①

(b) (i) These are weak intermolecular forces existing between all molecules. They are electrical in nature and occur in all polar molecules. ✓✓ ✗✗ ②

  (ii) An example is that the forces are present in liquid HI. ✓ ✗ ③

(c) The stronger the van der Waals force the higher the boiling temperature of a liquid. ✓ ④

Examiner commentary

① Correct

② Minimal answer but no mention of induced dipole–induced dipole.

③ One example only.

④ OK

Tom is awarded 5 marks out of 8: Grade C.

Seren's answer

(a) Van der Waals ✓

(b) (i) Weak intermolecular forces that exist between all atoms and molecules. They are electrical in nature and originate through the attraction between opposite charges. In one type the molecules are polar and have permanent charge separation. In the second type, present in all molecules, fluctuating dipoles induced by electron movements come into phase to give an attractive force between the molecules. ✓✓✓✓

  (ii) An example of the first, dipole–dipole, type is HI and of the second, induced dipole–induced dipole type is He. ✓✓

(c) The boiling temperature of a liquid is governed by the strength of the van der Waals force between the molecules. ✓

Examiner commentary

Good answer with part (b) showing excellent understanding.

Seren is awarded 8 marks out of 8: Grade A.

Q & A

14

Hydrogen bonding

Explain the nature of the hydrogen bond [4] and describe and explain its effect on the boiling temperatures of liquids containing hydrogen bonds [2] and on the solubility of compounds in water. *[2].*

Tom's answer

The hydrogen bond is an intermolecular bond where hydrogen bonded to an electronegative element such as N, O or F bonds to a similar element in another molecule.✓✓✓ ①

It is a very strong bond. ✗ ②

The boiling temperatures of liquids having hydrogen bonding are higher than expected ✓③ because the molecules in the liquid are held together more strongly so that more energy (i.e. a higher temperature) is needed to separate them.✓ ④

Examiner commentary

①&② Correct, except that the H bond is only relatively strong compared with van der Waals bonding but is much weaker than covalent, ionic and metallic bonds

③&④ Boiling temperatures sound

✗ No mention of solubility

Tom achieves 5 out of 8 marks: Grade C

Seren's answer

The hydrogen bond is an intermolecular bond where hydrogen bonded to an electronegative element such as N, O or F bonds to a similar element in another molecule. It is a stronger bond than a van der Waals bond but much weaker than a normal covalent bond. ✓✓✓✓

The boiling temperatures of liquids having hydrogen bonding are higher than expected because the molecules in the liquid are held together more strongly so that more energy (i.e. a higher temperature) is needed to separate them. ✓ ✓

Compounds such as lower alcohols having O-H bonds will be soluble in water through hydrogen bonding with the water. Many ionic solids are also soluble in water because of the interaction of the cation with the polar $O\delta$ - and the anion with the $H\delta+$ atoms in the water. ✓ ✓

Examiner commentary

A comprehensive answer. In the solubility part it is sensible to mention the solubility of ionic salts, although the interaction is perhaps not strictly hydrogen bonding in the normal sense.

Full marks are awarded: Grade A.

Shapes of molecules

(a) Complete the table below by inserting the numbers of bonding pairs of electrons and naming the shapes of the molecules involved. [4]

Molecule	No. bonding pairs	No. lone pairs	Shape
$BeCl_2$		0	Linear
PCl_3	3	1	
CCl_4		0	

(b) State and explain the difference between bond pairs and lone pairs of electrons in governing the shapes of molecules. [2]

(c) The bond angles in CH_4, NH_3 and H_2O are 109.5, 107 and 104.5 degrees respectively. Explain this. [3]

Tom's answer

(a) $BeCl_2$ 2 bond pairs ✓ ①
   PCl_3 trigonal planar shape ② ✗
   CCl_4 4 bond pairs, tetrahedral shape ✓ ✓ ③

(b) Lone pairs repel more around the central atom ✓ ④ because they have opposite spin. ✗ ⑤

(c) Methane has no lone pairs, ammonia has one and water has two so that the bond angle is squeezed down along the series. ✓ ✓ ✗ ⑥

Examiner commentary

② Correct in ① and ③ but there is a lone pair in PCl_3 as in ammonia that distorts the planar shape into a trigonal pyramid.

④ ⑤ Lone pairs do repel more than bond pairs but this has nothing to do with spin but only that lone pairs are closer to the central atom.

⑥ Factually correct but limited and needs an explanation of why the angle is reduced.

Tom has 6 marks out of 9: Grade B.

Seren's answer

(a) $BeCl_2$ 2 bond pairs
   PCl_3 trigonal pyramid shape
   CCl_4 4 bond pairs, tetrahedral shape ✓✓✓✓ ①

(b) Lone pairs have a larger repulsive effect around the central atom than bond pairs. This is because they are more localised around this atom while bond pairs are stretched out between the bonded atoms. ✓✓ ②

(c) Methane has four bond pairs only giving a symmetrical tetrahedral arrangement. In ammonia increased repulsion by the lone pair on the bond pairs closes up the H-N-H bond angle slightly and in water with two lone pairs the repulsion on the bond pairs is increased further giving a smaller H-O-H angle. ✓✓✓ ③

Examiner commentary

① All are correct.

② Good explanation that brings out the essential difference.

③ Again a good explanation.

Seren has shown a good understanding for the full 9 marks: Grade A.

Periodic Table

(a) (i) State the meaning of the term electronegativity, describe how its value changes across and down the Periodic Table and explain this trend. [3]

(ii) Explain why the melting temperature of magnesium (923 K) is greater than sodium (371 K). [1]

(iii) Explain why the melting temperatures of the halogen elements increase down the group. [2]

Tom's answer

(a) (i) A measure of the electron–attracting power of an atom in a covalent bond. ✓ ①
Its value increases across the Periodic Table and down a group. ✓ ✗②

(ii) Magnesium has two outer electrons to take part in bonding the solid but sodium only has one. ✓ ③

(iii) The van der Waals forces increase down the group as the number of electrons increases. ✓✗ ④

Examiner commentary

① & ② Partly correct but x decreases down a group not increases

③ Needs more explanation but just gets the mark.

④ The fact that we have covalent diatomic molecules needs to be stated first.

Tom has 4 marks out of 6: Grade B.

Seren's answer

(a) (i) A measure of the electron-attracting power of an atom in a covalent bond. Its value increases across the Periodic Table but decreases down a group, so effectively increasing diagonally upwards across the Table. ✓✓✓ ①

(ii) Magnesium has two outer electrons to take part in bonding the solid but sodium only has one. The electrons act as a 'glue' holding the metal atoms together and the more electrons per atom the stronger the binding and higher the m.t. ✓ ②

(iii) The halogens comprise covalent diatomic molecules held together in the solid by van der Waals forces. The van der Waals forces increase down the group as the number of electrons increases so that the m.t.s increase. ✓✓ ③

Examiner commentary

① Correct

② Good answer

③ Good. Seren notes that the strength of the vdW forces is proportional to the number of electrons in the atom.

Seren has the full 6 marks: Grade A.

Q & A 17

Carbon structures

(a) Explain in terms of bonding and structure, why diamond has a very high melting temperature. [2]

(b) Describe the structure and physical properties of carbon nanotubes. [2]

Tom's answer

(a) Each carbon atom is joined to four others by a strong covalent bond. ✓✗ ①

(b) These are tubes a few nm in diameter and much longer than their width that are made of graphite. ✓✗ ②

Examiner commentary

① The formation of an infinite 3D network needs to be stated.

② Just enough for one mark

Tom has 2 marks out of 4: Grade D.

Seren's answer

(a) Each carbon atom is joined to four others by a strong covalent bond and these form an infinite network in three dimensions. ✓✓ ①

(b) These are tubes a few nm in diameter and much longer than their width that are made as rolls of a graphite monolayer called graphene. They are very strong and their properties such as electrical conductance are different along the length of the tube compared with across the tube. ✓✓ ②

Examiner commentary

① Seren has this absolutely correct.

② Very extensive answers are possible but with only 2 marks available this is enough.

Seren has the full 4 marks: Grade A.

Periodic Table trends

(a) Describe and explain the general change in ionisation energies:
    (i) Across a period, e.g. from Na to Ar,
    (ii) Down a group, e.g. From Li to Cs.   [4]

(b) (i) Give one example of a (I) basic oxide, (II) acidic oxide.   [2]
    (ii) State the regions of the Periodic Table in which the elements form (I) basic and
        (II) acidic oxides.   [2]

Tom's answer

(a) (i) IEs increase across a period because there is a steady increase in the number of protons in the nucleus. ✓✓ ①

    (ii) IEs fall down a group because the outer electron is further from the nucleus in the larger atoms. ✓ ✗ ②

(b) (i) (I) MgO, (II) SO_2 ✓✓ ③

    (ii) (I) The LHS, (II) the RHS. ✓✗ ④

Examiner commentary

① Should have added 'without much increase in electron shielding'.

② This explanation itself is rather unsound (see Seren's answer for a better explanation of this), and the distance from the nucleus is a secondary factor controlled by the effective nuclear charge.

③ This is basically acceptable

④ Incorrect. He should have referred to regions of the LHS and RHS.

Tom scored 6 out of the 8 possible marks: Grade B.

Seren's answer

(a) (i) IEs increase across a period because there is a steady increase in the number of protons in the nucleus while the ionisable electron is in the same orbital and electron shielding is not much increased. ✓✓ ①

    (ii) IEs decrease slightly down a group since the increase in nuclear charge is outweighed by increased screening by the filled orbitals so that the effective nuclear charge decreases. ✓✓

(b) (i) (I) CaO, (II) NO_2 ✓✓

    (ii) (I) The LHS with the bottom of the s-block being most basic. ✓

    (II) The RHS, especially the upper parts of groups 5 and 6. ✓ ②

Examiner commentary

① Good answers showing clear understanding of the factors.

② Scores all the marks but the actual state of group 7 oxides is not in the spec. and inert gases do not form oxides.

Seren scored the full 8 marks: Grade A.

Oxidation numbers

(a) State four of the rules that are used to assign oxidation numbers to elements in compounds. [4]

(b) Evaluate the oxidation numbers of all of the atoms in the following compounds or ions:
$CaSO_4$, F_2, Na_2CO_3, NH_4^+ [4]

Tom's answer

(a) Elements are zero, oxygen is -2, hydrogen is 1, in ions the oxidation number is the charge on the ion. ✓✓✓✗ ①

(b) Ca is 2, O is -2, S is 6; F is 0, Na is 1, O is -2, C is 4; H is 1, N is 4 ✓✓✓✗ ②

Examiner commentary

① Three correct – can use 2 or +2 or II for positive numbers and -3 or -III for negative. The oxidation number is only the charge on the ion for atoms, such as +1 for Na^+ and not for ionic compounds. Do not write 2+ or 3- for oxidation numbers, these are for ionic charges only.

② All correct except for NH_4^+ where sum of N and all the H oxidation numbers must be +1 therefore N is -3 and H_4 is 4 times +1.

Tom gets 6 out of a possible 8 marks: Grade B.

Seren's answer

(a) Uncombined elements are 0, in simple ions the oxidation number is the charge on the ion, hydrogen is +1, the sum of the oxidation numbers in an ion equals the charge on the ion or in an uncharged compound equals 0. ✓✓✓ ①

(b) Ca is 2, O is -2, S is 6; F is 0, Na is 1, O is -2, C is 4; H is 1, N is +3 ✓✓✓✓ ②

Examiner commentary

①) All correct including the good point about the sums of the oxidation numbers.

② Correct, with the compound ion dealt with successfully.

Seren scores 8 out 8: Grade A.

s-Block elements

(a) Write a balanced equation for the reaction of calcium metal with hydrochloric acid. [1]

(b) State how group I and group II elements compare with regard to:
   (i) Their reactivity with water and oxygen
   (ii) The solubility of their salts. [4]

(c) State also how the solubilities of group II hydroxides and sulfates change in descending the group. [2]

Tom's answer

(a) $Ca + HCl = CaCl + H_2$ ✗ ①

(b) (i) Both groups react with water to give hydroxides or oxides and with oxygen to give oxides. ✓ ②
   (ii) All group I salts are soluble in water ✓ and all group II salts are insoluble. ✗ ③

(c) Hydroxides become more soluble down group II; sulfates become less soluble down group II. ✓✓ ④

Examiner commentary

① Incorrect; errors in balancing and valency are very common and basics must be memorised.

② OK but limited answer.

③ Some group II salts are soluble, such as nitrates and halides.

④ This is generally acceptable.

Tom scored 4 out of 7: Grade C.

Seren's answer

(a) Ca + 2HCl = CaCl$_2$ + H$_2$ ✓ ①

(b) (i) Both groups react with water to give hydroxides or oxides and with oxygen to give oxides. Group I elements are more reactive than those of group II and in both cases reactivity increases down the group. ✓✓

  (ii) All group I salts are soluble in water; in group II nitrates and halides are usually soluble, sulfates and hydroxides may or may not be soluble and carbonates are not very soluble. ✓✓②

(c) Hydroxides become more soluble down the group whereas sulfates become less soluble. ✓✓③

Examiner commentary

① Correct
② Good clear answers in both parts.
③ Correct
Seren has full marks: Grade A.

The halogens

(a) (i) State what is meant by a **displacement reaction** in the halogens.
   (ii) Explain why such reactions occur.
   (iii) Write a balanced equation for the reaction of chlorine with potassium iodide solution.
   [3]

(b) (i) State how you would determine whether a solution of a halide in dilute nitric acid contains chloride, bromide or iodide ions and what would be observed. *[2]*
   (ii) Explain why this test is useful in organic as well as inorganic chemistry and describe the additional steps needed on the organic case. *[2]*

Tom's answer

(a) (i) Reactions in which one halogen exchanges places with another halogen in a halide. ✓①

  (ii) Halogens higher in the Periodic Table are stronger oxidising agents. ✓

  (iii) Cl$_2$ + 2KI = I$_2$ + 2KCl ✓ ②

(b) (i) Add silver nitrate solution when a precipitate would form. ✓ ✗ ③

  (ii) To identify the halogen in the organic compound: ✓ the organic halide must be heated. ✗ ④

Examiner commentary

① Correct but (i) and (ii) rather terse
② OK
③ Nothing about distinguishing the halogens
④ Nothing useful in the second part here

Tom scored 5 out of 7: Grade B.

Seren's answer

(a) (i) Reactions in which one halogen exchanges places with a lower halogen in a halide.

  (ii) Halogens higher in the Periodic Table are stronger oxidising agents and thus remove an electron from the halide to liberate the free halogen while being reduced to the halide.

  (iii) Cl$_2$ + 2KI = I$_2$ + 2KCl ✓✓✓

(b) (i) Add silver nitrate solution when a precipitate of silver halide forms. Silver chloride only dissolves in dilute ammonia, identifying chloride and silver iodide is yellow in colour as against the buff colour of bromide. ✓✓

  (ii) This test is useful in identifying the halogen present in halogenoalkanes but the halogen must first be liberated from the organic compound by warming with NaOH solution to hydrolyse it and liberate the halide ion. The solution must then be acidified with nitric acid so that only silver halide is precipitated when silver nitrate is added. ✓✓

Examiner commentary

Very clear and confident answers here with some additional material to impress the examiner.

Seren obtained full marks: Grade A.

Spectra and intermolecular forces

Ethanoic acid, CH_3COOH, commonly known as acetic acid, is an organic acid that gives vinegar its sour taste and pungent smell.

(a) Ethanoic acid contains C–O, C=O and O–H bonds and has the infrared spectrum shown below. Using the data sheet, label the characteristic absorptions for each of these **three** bonds on the spectrum. [2]

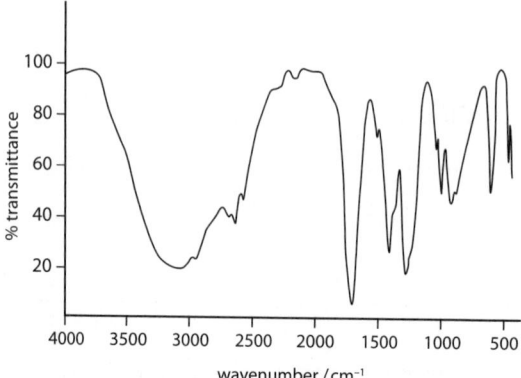

(b) The mass spectrum of ethanoic acid is shown below.
Explain how this shows that the formula for ethanoic acid is CH_3COOH. [2]

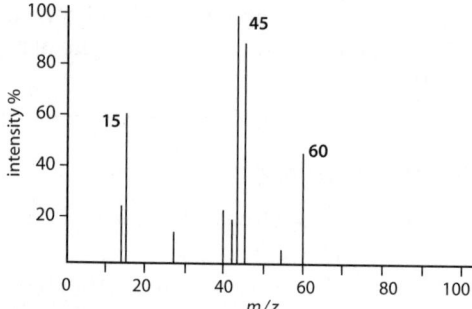

(c) The crystal structure of ethanoic acid shows that the molecules are found in pairs with hydrogen bonds between each pair.
Complete the diagram to show how two molecules of CH_3COOH can join together through hydrogen bonding. [1]

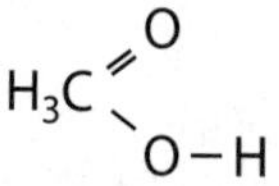

$$H_3C \underset{O-H}{\overset{O}{\diagdown}}$$

(d) Ethanoic acid can be formed from the oxidation of ethanol by potassium dichromate (VI).
   (i) State the conditions needed for this reaction to take place. [1]
   (ii) State what you would observe during the reaction. [1]

(e) The boiling temperature of ethanol is 78°C. Giving a reason in both cases, state how you would expect the boiling temperatures of the following compounds to differ from that of ethanol. [2]
   (i) Propane   (ii) Butan-1-ol

[Total: 9]

Tom's answer

(a) Peak at about 2800 to 3200 labelled OH, peak at about 1700 labelled C=O peak at about 1250 labelled C-O. ✓✓ ①

(b) M_r is 60 M_r ethanoic acid also 60 ✓ ✗ ②

(c)

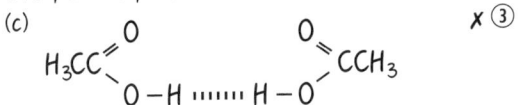

✗ ③

(d) (i) Heat ✗ ④

(ii) Colour change from orange to green ✓

(e) (i) Propane: lower boiling temperature than ethanol ✗ ⑤

Butan-1-ol: higher boiling temperature than ethanol. ✗

Examiner commentary

① Three labels correct

② The comment on the M_r is correct but more evidence i.e. on the nature of some of the fragments is needed to confirm that it is ethanoic acid.

③ Tom did not realise that hydrogen bonds are between H and O.

④ Although 'warm/heat' is a correct condition for many organic reactions, in this case the dichromate (VI) must be acidified.

⑤ Although whether the boiling temperature is higher or lower has to be correctly stated, in this type of question the marks are for the explanations. That is why Tom has not gained any marks here.

Tom achieves 4 out of 9: Grade C/D borderline.

Seren's answer

(a) Peak at about 2800 to 3200 labelled OH, peak at about 1700 labelled C=O. ✓ ✗ ①

(b) Peak at 60 is M_r and M_r of ethanoic acid is 60 ✓ Other peaks are fragments and one at 15 shows molecule contains CH_3. ✓ ②

(c)

$$H_3CC \overset{O\, \cdots\cdots\, H-O}{\underset{O-H\, \cdots\cdots\, O}{<\qquad>}} CCH_3$$

✓ ③

(d) (i) Reflux ✗ ④

(ii From orange to green ✓

(e) (i) Propane boiling temperature is less because propane has no hydrogen bonds but ethanol does. ✓ ⑤

(ii) Butan-1-ol boiling temperature is higher because it has a longer carbon chain. ✗

Examiner commentary

① Bold type is used to highlight important instructions/ information in the question that you might otherwise overlook, as Seren did here missing out the third bond needed.

② The two marks available suggest that two pieces are information are needed and Seren did this. She could also have commented that the peak at 45 was due to COOH⁺.

③ .... is an acceptable symbol for a hydrogen bond.

④ 'Reflux' is insufficient since the dichromate (VI) must be acidified.

⑤ Seren was awarded the first mark although it would have been more secure if she had also mentioned that propane only has Van der Waals forces. To score the second mark she should have gone on to state that the longer carbon chain meant that stronger Van der Waals forces were therefore possible in butan-1-ol.

Seren achieves 6 out of 9: Grade B.

Hydrocarbons and *E-Z* isomerism

Q & A 23

(a) Gas oil is a hydrocarbon fraction obtained from petroleum.
  (i) State how gas oil and other hydrocarbon fractions are obtained, starting from petroleum. [1]
  (ii) State why some of the gas oil fraction is cracked. [1]

(b) Tridecane, $C_{13}H_{28}$, is one of the compounds present in gas oil.
  One of the equations used to represent the cracking of tridecane is shown below.

$$C_{13}H_{28} \rightarrow \text{Compound Z} + C_4H_6 + H_2$$

  (i) Find the molecular formula of compound Z by using the equation. [1]
  (ii) Write the molecular formula of a compound which is in the same homologous series as compound Z but contains **six** carbon atoms per molecule. [1]

(c) Another of the products made by cracking tridecane is but-1,3-diene.

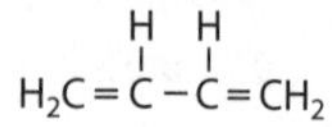

$$H_2C = C - C = CH_2$$
(with H, H on the two central carbons)

But-1,3-diene reacts with bromine to form several products.

  (i) One of the products is 3,4-dibromobut-1-ene, $CH_2=CH-CHBr-CH_2Br$.
  A possible mechanism for this bromination is shown below.

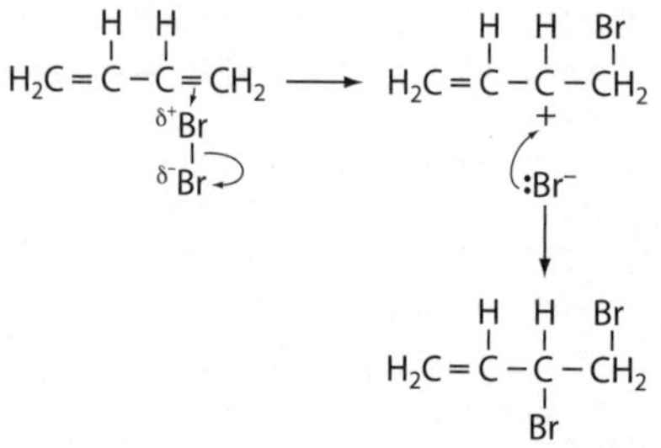

  I   State what is represented by the curly arrow. [1]
  II State what is represented by the δ+ and δ- symbols on the bromine atoms. [1]
  III The mechanism shows the formation of a carbocation **A**.
     Explain why the mechanism is less likely to proceed via carbocation **B**. [1]

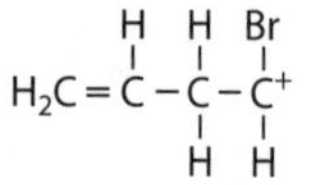

carbocation B

  (ii) Another product of the bromination of but-1,3-diene is 1,4-dibromobut-2-ene, $BrCH_2-CH=CH-CH_2Br$.
  This shows E-Z isomerism.
  I   Draw the structure of 1,4-dibromobut-2-ene to show the Z isomer. [1]
  II Explain why 1,4-dibromobut-2-ene shows E-Z isomerism. [1]

[Total: 9]

Tom's answer

(a) (i) Distillation. ✗ ①

(ii) Cracked to make smaller, more useful hydrocarbons. ✓②

(b) (i) C_9H_{18} ✗ ③

(ii) C_6H_{12} ✓

(c) (i) I Curly arrows show electron movement. ✗ ④

II The δ+ and δ− show the presence of a dipole. ✓

III Carbocation A is more stable. ✗⑤

(ii) ✓ ⑥

$$BrCH_2 \diagdown C = C \diagup CH_2Br$$
$$H \diagup \qquad \diagdown H$$

II The compound has a double bond. ✗⑦

___

Examiner commentary

① 'Distillation' alone is never acceptable when the answer' fractional distillation' is required.

② An acceptable answer. Tom does state that the molecules are smaller and why this is important.

③ Tom actually makes a mistake here in the formula of compound Z. However, once he has given the formula of an alkene in (i) then it follows that the formula in (ii) should also be an alkene. This is an 'error carried forward' and so he scores the mark for (ii).

④ This is a good example of why precise terminology is needed. It is the actual movement of a **pair** of electrons that are shown by the curly arrow. Tom's answer is not wrong but is imprecise.

⑤ As in I, it is the detail that means Tom's answer does not score any marks. What is there about the carbocation A that makes it more stable? He could alternatively have said that it is a secondary carbocation.

⑥ Tom is correct. It's a good idea if you can think of some hint or tip that shows you a way of remembering which is which!

⑦ Once again Tom's answer is not wrong but it does not score since it does not explain the significance of the double bond in the context of the existence of the two isomers.

Tom scores 4 out of 9 marks: Grade C

Seren's answer

(a) (i) Fractional distillation – separates according to the different boiling points of the fractions. ✓ ①

(ii) It creates smaller molecules that can be used, for example, in petrol. ✓ ②

(b) (i) C_9H_{20} ✓

(ii) C_6H_{14} ✓

(c) (i) I The curly arrow shows the movement of a pair of electrons. ✓ ③

II δ+ shows that end of the molecule is slightly positive and δ- shows that end of the molecule is slightly negative. ✓

III Carbocation A is more stable because it has 2 C atoms attached to the C atom with the +. ✓

(ii)

$$H \diagdown C = C \diagup CH_2Br \qquad ✗ ④$$
$$BrCH_2 \diagup \qquad \diagdown H$$

II There is no rotation about a carbon to carbon double bond. ✓

___

Examiner commentary

① From the wording of question, the answer 'fractional distillation' alone might have been sufficient but Seren makes sure of the mark by explaining the meaning of the term.

② Although Seren takes a different approach to Tom here, both answers are acceptable – both highlight the fact that the molecules are smaller and both explain, in a different way, the significance of this.

③ Seren accurately describes the type of movement of the pair of electrons that are shown by the curly arrow.

④ Incorrect – Seren has drawn the E form.

Seren scores 8 out of 9 marks: Grade A

Q &A

24

Calculation of formulae and polymers

(a) Compound **A** contains carbon, hydrogen and oxygen only. It has a molar mass of 88.2g mol^{-1}. Quantitative analysis of the compound shows that its percentage composition by mass contains 54.5% carbon and 9.10% hydrogen.

Calculate both the empirical and molecular formulae of compound **A**. *[4]*

(b) (i) Propan-1-ol can be completely oxidised to form compound **B**.

Name compound **B** and write the equation to show this oxidation. You may use [O] to represent the oxidising agent. *[3]*

(ii) Propan-1-ol can also form propene by a dehydration reaction. Name a suitable reagent for this reaction. *[1]*

(c) Propene can be polymerised to form poly(propene). Give the formula of the repeating unit in poly(propene). *[1]*

(d) Substituted alkenes can also be polymerised to give useful polymers. Name an important polymer formed from a substituted alkene. *[1]*

[Total 10]

Tom's answer

(a) % oxygen = 36.4 ✓

$$C : H : O$$

$$= \quad \frac{54.5}{12} : \frac{9.10}{1} : \frac{36.4}{16}$$

$$= \quad 4.54 : 9.01 : 2.28 ✓$$

$$= \quad 5 : 9 : 2$$

Empirical formula = $C_5H_9O_2$ ✗

Empirical M_r = 101

Molecular formula = ✗ ①

(b) (i) Propanoic acid ✓

$C_3H_7OH + [O] \rightarrow C_3H_7COOH$ ✗ ✗ ②

(ii) sulfuric acid ✓ ③

(c)

$$-\overset{\overset{\displaystyle CH_3}{|}}{\underset{\underset{\displaystyle H}{|}}{C}}-\overset{\overset{\displaystyle H}{|}}{\underset{\underset{\displaystyle H}{|}}{C}}-H \qquad ✗ ④$$

(d) PVC ✓ ⑤

Examiner commentary

① Tom uses the percentages to calculate the number of moles of each element, but he then approximates these to whole numbers. In exam questions the numbers will always produce obvious whole numbers at this stage and you should never approximate. Tom might have realised he had made a mistake, and gone back and corrected it, when his empirical formula M_r was not related directly to the value given in the question.

② Tom named the acid correctly but did not recognise that one of the carbons in the C_3H_7 group is needed to form the COOH in the acid.

③ A variety of acceptable dehydrating agents exist. Tom's answer, sulfuric acid, was accepted although concentrated sulfuric acid would have been better.

④ In drawing the formula for the repeat unit of a polymer the end bonds must show that the chain continues, i.e. have nothing attached to them.

⑤ There are generally a wide variety of answers to questions that ask for names or uses of particular groups of compounds. It is important to note **important** and not quote lab type/ small scale uses.

Tom scores 5 out of 10: Grade C.

Seren's answer

(a) % oxygen = 36.4 ✓

$$C : H : O$$

$$= \quad \frac{54.5}{12} : \frac{9.10}{1} : \frac{36.4}{16}$$

$$= \quad 4.54 : 9.01 : 2.28 ✓$$

$$= \quad 1.99 : 3.95 : 1$$

Empirical formula = C_2H_4O ✓

Molecular formula = $C_4H_8O_2$ ✓ ①

(b) (i) Propanoic acid ✓

$C_3H_7OH + [O] \rightarrow C_2H_5COOH + H_2O$ ✓ ✗ ②

(ii) Aluminium oxide ✓

(c)

$$-\left[\overset{\overset{\displaystyle CH_3}{|}}{\underset{\underset{\displaystyle H}{|}}{C}}-\overset{\overset{\displaystyle H}{|}}{\underset{\underset{\displaystyle H}{|}}{C}}\right]- \qquad ✓ ③$$

(d) PTFE ✓ ④

Examiner commentary

① Seren correctly uses the percentages to calculate the number of moles of each element, and then divides by the smallest to obtain a whole number ratio. Although she scored full marks, she might have been more secure in this if she had shown her calculation of the M_r of her empirical formula.

② Seren named the acid correctly and knew that one of the carbons in the C_3H_7 group is needed to form the COOH in the acid. While she scored the mark for this and realised that water was the other product, she did not actually balance the equation.

③ The end bond shows that the chain continues, i.e. has nothing attached.

④ One of a large variety of acceptable answers.

Seren scores 9 out of 10: Grade A.

Halogenoalkanes

(a) Methane reacts with gaseous chlorine giving chloromethane and hydrogen chloride.

$$CH_4(g) + Cl_2(g) \rightarrow CH_3Cl(g) + HCl(g)$$

In a report of this reaction, a student came across a number of terms.
Illustrating your answer with an equation in **each** case, state what is meant by:

(i) homolytic fission, [2]

(ii) a propagation stage. [2]

(b) One of the products of the reaction between ethane and chlorine is 1,1,1-trichloroethane.

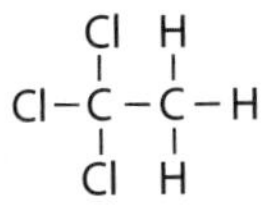

$$\begin{array}{ccc} Cl & H & \\ | & | & \\ Cl-C-C-H & \\ | & | & \\ Cl & H & \end{array}$$

The manufacture and use of 1,1,1-trichloroethane is now restricted because of its adverse effects on the ozone layer. However, the corresponding fluorocompound 1,1,1-trifluoroethane does not cause environmental problems in the ozone layer.

(i) Explain why only the chloro-compound has these adverse effects. [2]

(ii) A sample of 1,1,1-trichloroethane is reacted with an excess of sodium hydroxide solution and then acidified.

I   One of the products of this reaction is liquid R whose mass spectrum shows a molecular ion at m/z 60.

The infrared spectrum of R shows characteristic absorption frequencies at 1750 cm^{-1} and 2500-3500 cm^{-1}.

Use this information, showing your working, to suggest a structural formula for liquid R. [4]

II Chloride ions are also produced when 1,1,1-trichloroethane reacts with aqueous sodium hydroxide. The products of the reaction are then acidified with nitric acid and the mixture tested for the presence of chloride ions.

State the reagent(s) used and the observations when the mixture was tested for chloride ions. [2]

[Total 12]

Tom's answer

(a) (i) Homolytic fission is when a bond breaks and each atom receives an electron. ✗ ①
$Cl-Cl \rightarrow 2Cl^\bullet$ ✓

(ii) A radical reacts and another is formed to carry on the reaction. ✓
$Cl^\bullet + CH_4 \rightarrow CH_3^\bullet + HCl$ ✓

(b) (i) The C–F bond is strong ✗
and is not broken in the ozone layer. ✗ ②

(ii) I $M_r = 60$ ✓
Infrared peaks show C=O ✗ O–H. ✗
Liquid R is ethanoic acid. ✗ ③

II The reagent used is aqueous silver nitrate ✓
and during the reaction a white colour is seen. ✗ ④

Seren's answer

(a) (i) Homolytic fission is where a covalent bond breaks in an organic molecule. ✗ ①
$Cl-Cl \rightarrow Cl^\bullet + Cl^\bullet$ ✓

(ii) In a propagation stage, a radical takes part and regenerates another. ✓
$CH_3^\bullet + Cl_2 \rightarrow Cl^\bullet + CH_3Cl$ ✓ ②

(b) (i) Only the chlorocompound has adverse effects because the C–F bond is stronger than the C–Cl bond ✓ so that it is not broken by uv radiation. ✓

(ii) I The m/z peak quoted shows that the M_r is 60 ✓
In the IR spectrum the peak at 1750 cm^{-1} shows the presence of C=O ✓ and the one at 2500 to 3500 cm^{-1} shows O–H ✓
This means R is ethanoic acid CH_3COOH. ✓

II $AgNO_3$ is used ✓ and this gives a white precipitate. ✓ ③

Examiner commentary

① Tom does not score the first mark since two points were needed. The bond broken must be covalent and each of the joined atoms must receive one of the electrons.

② When a question includes words such as only/ compare/ difference between, a comparison of some sort is required. In this case it was necessary to explain the difference in terms of the strengths of the carbon to halogen bond.

Tom's answer does not include what causes the bond to break.

③ A mass spectrum is nearly always used to give the M_r and an infrared spectrum to show the bonds, and therefore functional groups, present. Tom appreciates this but the absorption frequencies that indicate a particular bond must be specified. Tom realises that R is ethanoic acid but the question states that a formula must be given.

④ Observations must include colour and if a solid precipitate is formed.

Tom scores 5 out of 12: Grade D.

Examiner commentary

① Seren's answer does not receive the first mark either for the same reasons as Tom. Note that Seren used a different equation from Tom but both are acceptable to show homolytic fission.

② Although different from Tom's answer, both equations and definitions are acceptable.

③ Since the question says 'state the reagents', names or formulae are acceptable. When a question asks for an observation and a solid is formed, answers must specify this and its colour.

Seren scores 11 out of 12: Grade A.

Alkenes

Compound A can be converted to 2-bromobut-2-ene in two steps:

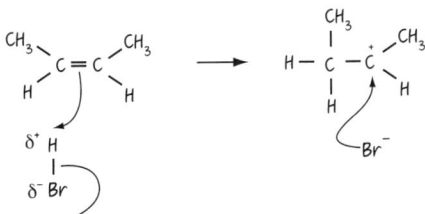

compound A   compound B

(a) During step 1, compound A is bubbled through bromine water to produce a layer of compound B.
   (i)   Give the colour change that would be noted during step 1.   *[1]*
   (ii) **Name** compound B.   *[1]*
   (iii) Step 2 is performed using similar reagents and conditions to those used in the production   of ethene from bromoethane. Give the reagents and conditions required for this reaction.   *[2]*
(b) (i)   Compound A also reacts with hydrogen bromide, HBr. Give the mechanism for this reaction.   *[4]*
   (ii) Classify the mechanism of the reaction in (b) (i).   *[1]*
                                                                                    [Total 9]

Tom's answer

(a) (i) It turns colourless. ✗ ①
   (ii) 2-dibromobutane. ✗ ②
   (iii) Sodium hydroxide ✓, dissolved in ethanol ✗ ③
(b) (i)

dipoles ✓ arrows ✗ structure of carbocation ✓ arrow from Br⁻ ✗ ④
   (ii) Addition ✗ ⑤

Examiner commentary

① When a question asks for the colour change, or observations, it is essential to include the colour at the start and end of the reaction.

② In the name of a disubstituted compound the positions of both substituents must be specified.

③ Many questions ask for reagents and the conditions needed for a reaction. Tom states a correct reagent, an alkali, and recognises that it must be dissolved in ethanol. However, heating/reflux is an essential condition. This is true in many reactions in organic chemistry!

④ The dipoles and first arrow are correct.

However, the second arrow starts correctly from the bond but does not go to the bromine.

The structures of the carbocation are acceptable – the + is quite clearly on the correct carbon.

A curly arrow in a mechanism shows the movement of a **pair** of electrons. Tom's final arrow does not start from a lone pair.

⑤ 'Addition' is not wrong but, in classifying a reaction, the nature of the initial attack should be stated (as well as the overall effect of the reaction).

Tom scores 3 out of 9: Grade D.

Seren's answer

(a) (i) Colour change is brown to colourless ✓ ①

(ii) 2,3-dibromobutane ✓

(iii) Sodium hydroxide ✓ dissolved in ethanol and then refluxed with compound B. ✓

(b) (i)

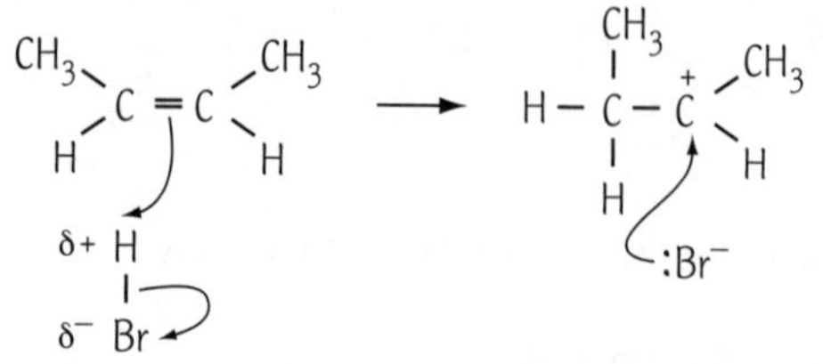

dipoles ✓ arrows ✓ structure of carbocation ✓ arrow from lone pair ✓ ②

(ii) Electrophilic addition ✓ ③

Examiner commentary

① The initial and final colours are quoted.

② It is really important that arrows start and finish in the correct places. They generally start at a bond or a lone pair.

③ The types of initial attack you will see in this unit are radical, electrophilic and nucleophilic.

Quickfire answers

Topic 1

1. 35p, 35e, 44n and 35p, 35e, 46n.
2. (a) 11p, 10 e (b) 9p, 10e
3. Radioactive emissions are stopped by a lead shield thus preventing escape of the radioactivity.
4. ^{207}Tl
5. Radiation is ionising/releases high energy/produces radicals/breaks chemical bonds/causes cell mutation.

   This causes radiation burns/radiation sickness/cancer/leukaemia.
6. 24 days
7. Carbon-14 used in radio-dating/potassium-40 used to estimate the geological age of rocks/α emitters in smoke alarms/β emitters to regulate thickness of metal foil.
8. See below
9. (a) Increase, because nuclear charge increases steadily but there is not much change in shielding.

   (b) Decrease, because outer electron has increased shielding from inner electrons and it is further from the nucleus.
10. $Mg^+(g) \rightarrow Mg^{2+}(g) + e^-$
11. Group 2 because there is a large jump in energy between the 2nd and 3rd ionisation energy, therefore the third electron has been removed from a new shell.
12. (a) Line with 690 THz has the higher energy since $E \propto f$

    (b) Line with 460 THz has the higher wavelength since $f \propto 1/\lambda$.

13. In absorption spectra, energy is absorbed from light causing electrons to move from a lower energy level to a higher one. It is seen as dark lines against a bright background.

    In emission spectra, energy is emitted as electrons fall back from a higher energy level to a lower one. It is seen as coloured lines against a black background.
14. As frequency increases the energy levels in an atom get closer, therefore the energy difference between the levels decreases, so the lines get closer.
15. (a) 80.02 (b) 249.7
16. To prevent collisions between the sample and air molecules.
17. 87.7
18. The molecular ion splits to give Cl^+ ions.
19. 37.06 g
20. 0.2 mol dm^{-3}
21. 0.025

Topic 2

1. An equilibrium system where the forward and reverse reactions occur at the same rate.
2. (a) H_2SO_4 (b) NH_3 (c) KOH
3. It forms an H^+ ion.
4. 0.2 mol dm^{-3}
5. −1172 kJ mol^{-1}
6. −1250 kJ mol^{-1}
7. Measure the change in pressure at various times using a manometer. Measure the change in colour of NO_2 formation over time using a colorimeter.

8.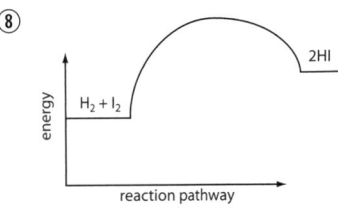

   130 kJ mol^{-1}
9. Fe in Haber process/V_2O_5 in Contact process/Ni in production of margarine.
10. Lower temperatures and pressures can be used so less CO_2 is formed during energy production.

    They are biodegradeable, therefore they can be easily disposed.

Topic 3

1. Equilibrium, kinetics and calculations or energetics.
2. The percentage of the total mass of the reactants contained in the desired product.

Topic 4

1. H $\overset{\circ}{\times}$ H

   H
   H $\overset{\times\circ}{\underset{\circ\times}{N}}$ H
   H

   Li $\overset{\times}{}$ $\circ\overset{\circ\circ}{\underset{\circ\circ}{Cl}}\circ$ $\longrightarrow$ Li$^+$ $\times\overset{\circ\circ}{\underset{\circ\circ}{Cl}}\circ^-$
2. F–F < H–Br < Br-F < O–H < Mg–O
3. The bonds between the strongly bonded N_2 units are weak van der Waals forces of the induced dipole-induced dipole type.
4. Van der Waals < hydrogen < covalent
5.

8. (a)(i)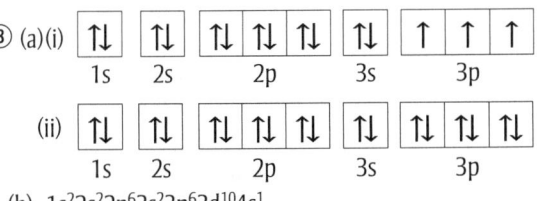

   (b) $1s^2 2s^2 2p^6 3s^2 3p^6 3d^{10} 4s^1$

⑥

```
    -lp-
   °o  o°
  o° S °+
   °+ °o
  H  Y  H
     bp
```

⑦ Linear, tetrahedral, tetrahedral and trigonal bipyramid respectively.

Topic 5

① Because the graphitic structure extends along the nanotube giving good graphitic electrical conduction and not between the nanotubes.

② A monolayer of the graphite structure giving a strong single sheet.

Topic 6

① B^{3+} and C^{4+}. Ionisation energies are increasing across the period and it is energetically unfavourable to remove three or four electrons successively.

② Across a period e.g. between Li and F electrons are being added to the same main shell so that there is little extra electron shielding to counteract the increased nuclear charge. However, the Na electron to be lost is screened by the full inner shell built up in the Li row and this offsets the extra protons in the nucleus.

③ (a) Na (0), Cl (0), $Na^+(+1)$ Cl^- (-1)

(b) -1

(c) Na (1), O(-2), S (6); N (5), F (-1); O(0); C (3), H (1)

④ (i) Yellow Na, brick red Ca, apple green Ba and lilac K.

(ii) (a) the hydroxides,
(b) the sulfates.

⑤ ½ Cl_2 (0) + K(1)Br(-1) = ½ Br_2(0) + K(1)Cl(-1)

Topic 7.1

① The longest carbon chain has 8 carbon atoms.
Name based on oct.

② (a)
```
      Cl H H
      |  | |      H
  H-C-C-C=C
      |  | |      \
      H  H         H
```

(b)
```
          H
          |
       H-C-H
       |       OH H H
       |       |  | |
  H-C--C---C-C-C-H
       |       |  | |
       H       H  H H
       |
    H-C-H
       |
       H
```

③ (a) 3-methyl pent-1-ene
(b) 2-bromo, 2-chloropropan-1-ol

④ (a) C_5H_9ClO

(b)
```
      H H H H
      | | | |    H
  H-C-C-C-C=C
      | | | |    \
      H Cl OH     H
```

(c)

⑤ Propanoic acid

⑥ C_nH_{2n}

⑦ $C_{72}H_{146}$

⑧ Empirical formula CH_2Br.
Molecular formula $C_2H_4Br_2$.

⑨ Pentane
```
      H H H H H
      | | | | |
  H-C-C-C-C-C-H
      | | | | |
      H H H H H
```

2-methylbutane
```
      H H H H
      | | | |
  H-C-C-C-C-H
      |   | |
      H   H H
      |
    H-C-H
      |
      H
```

2,2-dimethylpropane
```
          H
          |
       H-C-H
       H  |  H
       |  |  |
  H-C--C--C-H
       |  |  |
       H  |  H
       H-C-H
          |
          H
```

⑩ Pent-2-ene   Pent-1-ene
```
  /\/\         /\/\
```

⑪
```
  Cl        H
    \      /
     C=C
    /      \
  H         Br
```

⑫ Z isomer (Cl has a greater A_r than C)

Topic 7.2

⑬ Although it is actually 69°C you would need more information to be able to predict this. Any value reasonably above 36°C would be accepted (45°C to 100°C).

⑭ Pentane
```
      H H H H H
      | | | | |
  H-C-C-C-C-C-H
      | | | | |
      H H H H H
```

2,3-dimethylpropane
```
          H
          |
       H-C-H
       H  |  H
       |  |  |
  H-C--C--C-H
       |  |  |
       H H-C-H H
          |
          H
```

2,2-dimethylpropane is more branched than pentane. It therefore has a smaller surface area and less van der Waals forces. Since less intermolecular forces have to be broken to boil, the boiling temperature is less.

⑮ The sea creatures, etc., involved when the oil was formed are different so the composition of the petroleum would be different. Fractions would also be different.

⑯ The use for each fraction depends on its volatility and therefore its boiling temperature.

⑰ $C_{10}H_{22} \rightarrow C_6H_{14} + C_4H_8$ (note – the 2nd product is an alkene)

⑱ They are unreactive because they do not have charged areas to attract charged attacking species. They have no π bonds or dipoles.

⑲ Radicals are reactive since they contain an unpaired electron. It is more stable if the electron can gain another one to become paired.

⑳ A propagation reaction can form $CH_3^{\bullet}$. Two of these radicals then react.

$2CH_3^{\bullet} \rightarrow C_2H_6$

㉑ $CH_4 + 4Cl_2 \rightarrow CCl_4 + 4HCl$

㉒ When the bond breaks both electrons go to the atom on one side of the bond.

㉓ It is cheaper. Although catalysts should not be used up in reactions, some 'poisoning' is inevitable.

㉔ $(CH_3)_2C=CH_2 + HBr \rightarrow (CH_3)_2CBrCH_3$

㉕ Any suitable diagram including 3 carbon atoms joined to a +ve C.

㉖ Poly(1-chloro, 2-cyanoethene). It does not matter how complex the monomer is, the name of the polymer is just the monomer's name with poly in front.

㉗
```
   Cl  CN  Cl  CN
   |   |   |   |
 - C - C - C - C -
   |   |   |   |
   H   H   H   H
```

㉘ CH_2. In addition polymerisation nothing is lost and so the empirical formula of the polymer is the same as that of the monomer.

㉙
```
  Cl         CO2CH3
    \       /
     C = C
    /       \
  H           H
```

㉚ To lower the melting temperature van der Waals forces must be reduced. This could be done by making the polymer more branched or by lowering the M_r (and making the chain shorter).

Topic 7.3

㉛
```
   Cl  Cl  H   H
   |   |   |   |
 H-C - C - C - C-H
   |   |   |   |
   H   H   H   H
```

㉜

㉝ It is not a satisfactory method of preparation because it involves a radical substitution mechanism. A mixture of mono, di and polysubstituted products will be formed.

㉞ Heat organic compound with aqueous sodium hydroxide.

Neutralise excess sodium hydroxide with dilute nitric acid.

Add aqueous silver nitrate.

Add aqueous ammonia to the precipitate.

Result: if bromine present, cream precipitate with silver nitrate, soluble in concentrated aqueous ammonia.

㉟ $I^-(aq) + Ag^+(aq) \rightarrow AgI (s)$

㊱ Liquefied by putting them under a high pressure.

㊲ In a propagation step a radical is present at the start of the equation and another one is formed at the end.

㊳ $2O_3 \rightarrow 3O_2$. This equation shows the destruction of the ozone layer to produce oxygen.

Topic 7.4

㊴ Pentan-1-ol has a longer carbon chain and so more van der Waals forces are possible. These raise the boiling temperature.

㊵ Ethane only contains van der Waals forces which are weak and so its boiling temperature is low. Chloroethane is polar and so contains permanent dipole-permanent dipole attractions which are stronger – this raises the boiling temperature. Ethanol contains hydrogen bonds and, as these are the strongest of the intermolecular forces, its boiling temperature is the highest.

㊶ Ethene is formed when petroleum is cracked.

㊷ The mixture can be cooled so that ethanol condenses.

㊸ (a) $CH_3CH_2CH_2OH \rightarrow CH_3CH=CH_2 + H_2O$

(b) $CH_3CH_2CH_2OH + 2[O] \rightarrow CH_3CH_2COOH + H_2O$

㊹ A positive test is shown if the acidified dichromate(VI) changes from orange to green.

Topic 8

① Smaller quantities of materials are needed and the methods are less destructive of the sample.

② The relative abundance of each fragment. This is not useful in any spectrum you will be given – it really shows the stability of the fragments.

③ ^{35}Cl is about 3 times more abundant than ^{37}Cl. The peak heights would be in the ratio 3:1.

④ Peak at 60 (the M_r). The OH group is lost (m/z decreases by 17).

⑤ This is the range for C–H and this is present is nearly all organic molecules.

⑥ Absorption at 1720 cm^{-1} would not be present for CH_3OCH_3 since this does not contain C=O.